Ecology of Fish Pond

First Published, 2011
Reprinted, 2018

ISBN: 978-81-7035-739-1 (Harbound)

Published by : **Daya Publishing House®**
A Division of
Astral International Pvt. Ltd.
– ISO 9001:2008 Certified Company –
4736/23, Ansari Road, Darya Ganj
New Delhi - 110 002
Phone: 011-4354 9197, 2327 8134
E-mail: info@astralint.com
Website: www.astralint.com

Ecology of Fish Pond

– Authors –

Prof. D.R. Khanna

Dr. Rajani Rana

Dr. Gagan Matta

– Editors –

S.K. Pathak

Sunita Pathak

Mr. Vikas Singh

2018

Daya Publishing House®

A Division of

Astral International Pvt. Ltd.

New Delhi – 110 002

Preface

We can't imagine life on earth without water because it is such a unique component of nature which is having a vital role at micro to macro level in different life processes. Water as a medium is ideal for most of the living organisms on earth. Water availability on earth is abundant but most of it is in the form of salty water in oceans. Fresh water resources are limited and they are now continuously shrinking and degrading because of excessive exploitation by human beings.

From ancient times people were concerned about the water quality so methods to conserve fresh water resources were written in various Vedic as well as historic scriptures. Since the historic times, Kings as well as other renowned personalities were taken certain important steps to conserve and use these fresh water reservoirs sustainably. Those places where there was scarcity of water, people digged ponds, taalabs *etc.* to store rain water which otherwise went to oceans as run-off.

The pond ecosystem is a fresh water environment that can reveal the health of a local area. Specific life forms of fresh water environments such as the pond ecosystem show its overall health. Toxins or pollution can effect the pond ecosystem adversely. The importance of understanding the pond ecosystem involves the life forms and plant cultures that are part of the healthy environment. Agriculture and the impact of human activity on the pond ecosystem can effect the quality and purity of soil and water. Fertilizer, oil, introduced species, pollution; fishing can all upset the delicate balance that exists in a healthy pond ecosystem.

In this book, we have taken Paniyala pond which is a very much important source for nearby people of that area. This book presents an overview of pollution in pond ecosystem. Through this book we have tried to determine the present status of Paniyala pond, their conservation strategies *etc.*

Authors

Contents

	Preface	v
1.	Introduction	1
2.	Physico-Geographical Features	7
3.	Review of Literature	12
4.	Materials and Methods	19
5.	Parameters for Assessing Pond Water Quality	34
6.	Discussion	123
7.	Summary	149
	References	153

Chapter 1
Introduction

Water is unique component of nature having a very important role in the life of human beings. From industrial to agricultural and from commercial to domestic, without water we cannot imagine life on earth. Since time immemorable the great human civilization have originated, evolved and flourished around the water bodies. Water is of vital importance having properties, which are unique among all other liquids. The properties of water seem to be especially designed for the living organisms. Water covers more than 75 per cent area of the earth surface either as salt water or fresh water in which the fresh water is less than 2.6 per cent of the total. 4/5 of this fresh water is in the form of ice and thus useless for the mankind. The remaining amount really available is that in the form of surface and ground water.

Freshwater bodies are very essential for the existence of ecosystem contributing immensely in shaping and evolving the biotic and abiotic systems. The fresh water bodies can be classified into two categories *viz.* running water (Lotic) and standing water (Lentic). Reservoirs, Ponds, Lakes and Swamps come under second category; these are very useful for different activities besides drinking water sources. Freshwater is one of the scare natural resources and its conservation is assuming greater and greater significance due to ever increasing demand of water for drinking, irrigation and aquaculture. Ponds and reservoirs containing fresh water are used for fish culture, aquaculture, navigation and transport, recreation, irrigation hydropower generation and a host of such other purposes.

Water quality is primarily based upon its physical, chemical and biological characteristics. Over exploitation of these water bodies due to civilization have been causing deterioration of water quality. It is quite essential that the natural environment of the water body should be conducive to the extent that water should be used for drinking purpose; therefore besides considering limnological status, it is essential to monitor the quality of water. As water resources are limited and waste discharge becomes increasingly problematic, the concept of water reuse is becoming important. The global consumption of water is doubling every 20 yrs, more than twice the rate of human population growth. According to the United Nations, more than one billion people already lack access to fresh drinking water. If the current trend persists by 2025 the demand for fresh water is expected to rise by 56 per cent above the amount of water that is currently available.

The science, dealing with the study of physical, chemical and biological characteristics of inland water is termed as "LIMNOLOGY" which is derived from the Greek word 'Limnos' meaning pool, lake or swamp. Professor F.A. Forel who worked on Lake Geneva, Switzerland in 1892, 1895 and 1904 first introduced word, and he has been considered as the father of modern limnology. Welch (1952) defined limnology as "that branch of science which deals with the biological productivity of inland waters and with all the causal influences which determine it". In other words, limnology is the study of all types of fresh water bodies *viz* rivers, streams, lakes, reservoirs, ponds, pools, marshes, bogs and even under ground water. It deals with all the biotic and abiotic aspects of water. A suitable environment is necessary for any organism for proper exchange of essential substances between the organism and its surroundings. The chemical nature and physical conditions of water and presence of biotic flora and fauna have profound effect on the physiological process, growth performance, population density and geographical distribution of both animal and plant species.

Ponds are small (1.00 m^2 to about 5.00 ha), man-made or natural shallow waterbodies which permanently or temporarily hold water (De Meester *et al.*, 2005). They are numerous, typically outnumbering larger lakes by a ratio of about 100.00 to 1.00 (Oertli *et al.*, 2005), and occur in virtually all terrestrial environments, from polar deserts to tropical rainforests. Ponds are vital for a wide range of rare and endangered species. In countries where data are available, high concentrations of Red Data Book species are always found in ponds. Rare species are not only associated with the wet areas of ponds, but with the semi-aquatic margins.

Ponds are an important part of our culture: partly because of their intrinsic historical value, but also because their sediment records can tell us about our ancestors' way of life. Ponds are the 'local waterbody' and play a crucial role in maintaining and encouraging the link between people and wildlife. They also provide many opportunities for education and experimental research across a wide range of subjects. There is increasing evidence that ponds play a vital economic role in delivering ecosystem services: they offer sustainable solutions to support climate change amelioration and water resource management. They are also an asset for recreation and agriculture, particularly in the context of whole farm diversification and agri-tourism.

Ponds can contain water throughout the year or they can be seasonal, holding water for only part of the year. Ponds are valuable in many ways. They are also the sources of water for drinking, household and industrial requirements. Many plants and animals live in the water or at the water's edge or use ponds at a stage of their life cycle. They also provide source of water for livestock.

Pond ecology is best described as the interaction of the life in pond with the environment that exists there. A shallow, nutrient-rich pond exposed to sunlight with little water flowing through it will be teeming with algae and aquatic plants. It may have very little animal life present because of low oxygen levels. In contrast, a newly created, deep, spring-fed pond may have little life of any kind in it because of low temperatures and lack of food supply. Essential for the conservation of pond biodiversity is a good knowledge of its threats. Land use practices in the surroundings of ponds may, to an important extent, affect pond characteristics through a diversity of processes that play at the scale of the pond catchment (*e.g.* nutrient loading, increased erosion, pesticide contamination; Declerck *et al.*, (2005).

A pond begins with mostly water, few nutrients, and little aquatic life. Over time, the pond accumulates nutrients through an enrichment process called eutrophication. The addition of nutrients stimulates the growth of aquatic life. These organisms live, grow, and die. Their remains decay in the pond and the nutrients it took to grow them are released back into the water of the pond to keep the cycle going. Eventually, though, there will be an accumulation of material that resists decay and the pond will fill up. It will become a bog and someday will resemble dry land. The process of return to dry land can happen in a decade or may take centuries.

Despite their value, ponds are now amongst the most vulnerable and endangered of freshwater habitats. They face many threats, and receive little effective protection from legislation or policy. They serve as the most convenient and cheapest refuse dumping places for domestic and industrial wastes. So, water bodies are polluted and excessively fertilized. This tendency has resulted in indiscriminate and excessive loading of waste matter into the aquatic systems, beyond their capacity of self purification. The fresh water is becoming rare day by day as civilization draws nearer. Its quality has also equally deteriorated. Now man started realizing the effect of this pollution problem and trying to improve the quality of the water to reuse the same.

Aquatic life is influenced directly or indirectly by the physical, chemical and biological factors. Fluctuation in any one of the factors may create an adverse environment to the organisms, affecting their growth and life phenomena. All the hydrobiologists have given their attention towards it. The physico-chemical characteristics of any aquatic ecosystem and the nature and distribution of its biota are directly related to and influenced by each other and controlled by a multiplicity of natural regulatory mechanisms. However, because of men's exploitation of the water resources, the normal dynamic balance in the aquatic ecosystem is continuously disturbed, and often results in each dramatic response as depletion of fauna and flora, fish kill, change in physico- chemical characteristics etc. Artificial changes which lead to such ecological responses are referred to as pollution and pollutional stage may reach a stage when these valuable aquatic resources are no longer safe for human use.

The study of meteorological and physico-chemical conditions and their relationships with the flora and fauna in lakes, reservoirs and ponds is of great importance for the productivity of proteinacious food such as the fish etc. The morphological features of a lake basin that decisively affect the physico-chemical characteristics and productivity of the water bodies, still remains a neglected area of limnology in the Indian context. Physico-chemical characteristics are the most influencing parameters, which affect the life in water. Fluctuations in these constituents create an adverse environment to organisms and reduce their ability to compete with other population within the environment. The erosion and siltation has reduced the water holding capacity of pond and lakes and if they are not checked in time, the pond may be converted into a shallow lake. Turbidity caused by suspended silt (deposited silt), reduces photosynthesis which affects the food chain. Existing physico-chemical characteristics of the water are conductive to life of organisms. Depth, temperature, turbidity and light constitute the most important physical parameter on which the productivity of ponds depends. Turbidity caused by suspended silt significantly reduces photosynthesis, which affect the food chain. The eroded soil and other materials, which contains leachable nitrogen, phosphorus, sulphur and trace elements become the source of nutrients, which leads to eutrophication of the water body.

The role of planktonic organisms in an aquatic environment is an essential link in the food chain and they are capable of affecting the entire aquatic biota. A concentration of dissolved oxygen in water is required for organisms such as fish, tadpoles and many aquatic insects to live in a pond. The gills of these organisms extract the dissolved oxygen (O_2) from water and release carbon dioxide (CO_2). Sources of oxygen in water result from oxygen exchange between the atmosphere and the surface of the water through wind and wave action, oxygen derived from the photosynthetic action of submerged aquatic vegetation called macrophytes (macrophytes: Greek, macro = large; phyte = plant), and the algal bloom. The major source of oxygen in "healthy" ponds, and the primary source of oxygen in all warmwater ponds, is produced through the photosynthetic action of the phytoplankters.

The aquatic biology has been attempted by many workers with the aim to find out the taxonomic structure of various water bodies. A community in an ecosystem is not just a haphazard grouping of animals and plants, which live independently from each other, but is characteristics taxonomic composition with definite trophic relationship and metabolic patterns. In the determination of productivity plankters are key organisms but the information about the fresh water plankters organisms is scanty in India.

Natural eutrophication is a slow process of enrichment and is a part of ageing phenomenon. In nature eutrophication is a beneficial process in that it produces enhanced productivity in the water mass. However, man made accelerated eutrophication of inland water has been viewed as an environmental degradation by many workers Lund (1972), Vollenweider (1984). This invariably results in deterioration of water quality, which in many ways causes significant economic loss. Eutrophication in streams, lakes, ponds, reservoirs and the ocean, especially in protected areas like bays and lagoons, could adversely affect their aquatic life and use for recreation.

Nitrogen and phosphorus can come from many sources, including animal manures, organic fertilizers and commercial fertilizers. Nutrient pollutants can also move from various locations, including cropped fields, pastures and lawns in urban settings. Nitrogen in the nitrate form is highly soluble in water. Water moving through the root zone may wash nitrate downward, reaching tiles or drainage channels and potentially reaching groundwater or surface waters like river, pond, canal and reservoirs. However, during the construction of any pond no full proof methods have been advocated to prevent the eutrophication and siltation of the pond, it is only minimized up to a certain limit.

Effects of eutrophication include increased biomass of phytoplankters, shifts in phytoplankters to bloom-forming species which may be toxic or inedible, increased in blooms of gelatinous zooplankters, increased biomass of benthic and epiphytic algae, changes in macrophyte species composition and biomass, death of coral reefs and loss of coral reef communities, decreases in water transparency, taste, odor and water treatment problems, oxygen depletion, increased incident of fish kills, loss of desirable fish species, reductions in harvestable fish and shellfish, decrease in perceived esthetic value of the water body (Carpenter, 1928).

The physico-chemical quality of water bodies play an important role in the production of pisciculture. In the determination of productivity, morphometry of water body, biomass *etc.*, play a very important role. Proper management and exploitation may enhance the productivity of water which in turn is directly related with food chain and food web structure of water body. In the determination of productivity, plankters are key organisms. Information about fresh water planktonic organisms are scanty in India. The basic knowledge of fresh water plankters is very limited. There detailed study on biological and ecological relations is necessary.

In view of inland fish production in the world, the position of India is next to China and Russia. The total fish production is 47.89 lakh tones, contributed 20.97 lakh tones from inland and 26.92 lakh tones from marine sector respectively in the year 1994-95 (Pandey *et al.*, 1993). Thus fish production from inland sector has great significance. It contributes the major share in the protein rich food for domestic consumption. Badola and Singh (1981) reported fish fauna of Alaknanda River and in its tributaries. Khan and Siddiqui (1971) studied the primary productivity and seasonal changes in the tropical fish pond, Aligarh.

Inland fish production in the country witnesses a six fold increases from 0.22 million ton in 1950-51 to 1.30 million ton in 1987-88, the overall growth rate in India fish production is almost on a par with a global four fold increases. A quantum jump in inland fish production occurred during the 6th and 7th plan period which witnesses overall growth rate more than 5.00 percent per annum against a declining trend noticed in the world as a whole. The country has 7.30 lakh hectares of ponds and tanks that can be used for fresh water aquaculture, though only 1.50 lakh hectares are being utilized at present. The remaining 6 lakh hectares either remain fallow or produce fish at subsistence level.

Freshwater bodies are important landscape and ecological resource, and are valued for recreation activities. Some activities on the surface of water bodies, or land adjoining water bodies, have few adverse effects. Other activities have potential to compromise important water values, by generating effects on amenity, cultural and ecological values, and on water quality and quantity. Past extractions, and modifications to these water bodies have already significantly altered their natural character. Water bodies include springs and groundwater, both of which may become polluted from various land use activities. Safeguarding the life supporting capacity of water is an important part of sustainable management and is of cultural importance to our country. Freshwater resources are reduced further by pollution. Reliable data on freshwater pollution is not available but according to one estimate global wastewater production is about 1,500.00 km^3. So one liter of wastewater pollutes eight liters of freshwater and thus present burden of pollution may be up to 12,000.00 km^3 worldwide.

Evaluation of water requirement is essential for all human activities, ranging from drinking to agricultural production and industrial development to all forms of large- scale energy generation. Accordingly, as the total global population increases, so do the aggregated human activities, which would increase water requirements (Biswas, 1998). This contributes to two contradictory trends, which further complicate the water-management process. On one hand a country's water requirements steadily increase with higher levels of human activities: on the other hand per capita water available declines steadily since the total amount of freshwater available is limited.

In rural area water bodies are mainly used for agricultural and drinking purposes. Application of excessive fertilizers in agriculture sector, irrigation and livestock bathing are putting a great strain on availability of fresh water bodies in rural areas. Urban water bodies are used for a variety of purposes, including: (1) potable and non-potable water supply; (2) disposal of urban runoff and overflows from sewer systems; (3) aesthetic enjoyment by the urban population; (4) aquatic life propagation; (5) contact and non-contact recreation (sailing and fishing); (6) flood conveyance; (7) other uses that may include cooling, navigation and groundwater recharge. Anthropogenic and biocentrism interests that may conflict drive the uses of the urban water bodies; therefore, the conflicting uses must be optimized (Eggleston, 2003).

Our country is bestowed with bountiful natural resources in the form of extensive coast line, river systems, estuaries, ponds, tanks, lakes, reservoirs and bells, exposed to the warm, tropical climate, these water bodies are extremely productive and they harbour enviable spectrum of fish genetic resources. Optimum utilization of these resources lead to many fold increase in inland fish production earning the country a place among the top in land fish producing nations of the world.

Therefore for scientific utilization of water resources, the study of phytoplankters and physico-chemical characteristics of water is of primary interest. With the same concern the present study was done in Paniyala Fish Pond, Roorkee in district Haridwar (India).

Chapter 2

Physico-Geographical Features

Roorkee, a moderate size town in the district of Haridwar (India), located at the bank of Ganga Canal, which takes off at Haridwar, 30.00 km away. About 30.00 km South of Shivaliks and about 180.00 km North of Delhi, it is linked by rail to major parts of the country such as Delhi, Mumbai, Kolkata, and is also connected with the roads. The nearest airport is about 65.00 km away at Dehradun.

The rain fall mainly takes between July and September, the average being 1050.00 mm. Besides, the University, which has 150.00 hectare (approx.) campus, the town is an important center for engineering, the Central Building Research Institute and the U.P. Irrigation Research Institute with the activity of Bengal Engineers, National Institute of Hydrology, U.P. Irrigation Design Organization being located here. The town has a total population of approx. 1,25,000.00 and is surrounded by Haridwar, Saharanpur and Muzaffarnagar Districts.

Pond is situated at the Village Paniyala, which is 7.00 km away from Roorkee bus stand and 4.00 km from Roorkee Railway station. At the western bank of the pond Paniyala Village is situated, and at the eastern bank buildings of Fisheries Department are there. At the southern side also the Paniyala village is situated and at the northern side the village gardens are present.

The source of water in the pond is rain water and quite a bit from ground water. The Paniyala pond is at height of 268.00 meter from the mean sea level (Longitude 77° 53″ E and Latitude 29° 51″ N). The maximum depth of pond is 2.80 meter and minimum 0.60 meter, mean depth is 1.70 meter. The surface area of the pond is 2,92,500.00 square meter, catchments area is 2,92,400.00 square meter (Table 2.1 and Map 2.1). Pond receives little domestic sewage from drains of adjacent villages, and in rainy season fertilizers, pesticides are also swept with rain water and mixes in pond. The study was conducted during 2007-2009 with regular monthly sampling of all the four sampling sites of pond.

Table 2.1: Physico-Geographical Features of Paniyala Pond, Roorkee

Sl.No.	Parameters	Measurements
1.	Height (from sea level)	268.00 meter
2.	Latitude	29° 51" N
3.	Longitude	77° 53" E
4.	Length of Pond	400.00 meter
5.	Width of Pond	250.00 meter
6.	Minimum depth	0.60 meter
7.	Maximum depth	2.80 meter
8.	Mean Depth	1.70
9.	Surface Area	2,92,500.00 square meter
10.	Catchments Area	2,92,400.00 square meter

Sampling Sites

Four sampling sites were selected for the collection of sample for physico-chemical and biological studies of the reservoir (Figure 2.1 to Figure 2.4).

Site–I

Sampling Site–I is situated at the Eastern bank of the pond near the building of Fishery Department. The depth of water is 1.50 meter. This sector receives a little bit of waste water from near by area (Figure 2.1).

Site–II

Sampling Site–II is situated at the Southern bank of the pond. This sampling site is near to the Paniyala village and depth of water at this site is 0.60-0.75 meter. This sector receives little sewage from village drains. This site is also used by some villages for washing their tractors and bullock carts (Figure 2.2).

Site–III

Sampling Site–III is situated at the Western bank of the pond. This sampling site is also near the Paniyala village and depth of water at this site is 2.00 m. (Figure 2.3).

Site–IV

Sampling Site–IV is situated at the Northern bank of the pond. This sampling site is near the Gardens of Paniyala village and depth of water at this site is 2.25 meter (Figure 2.4).

The bed of pond in this sector is clayey.

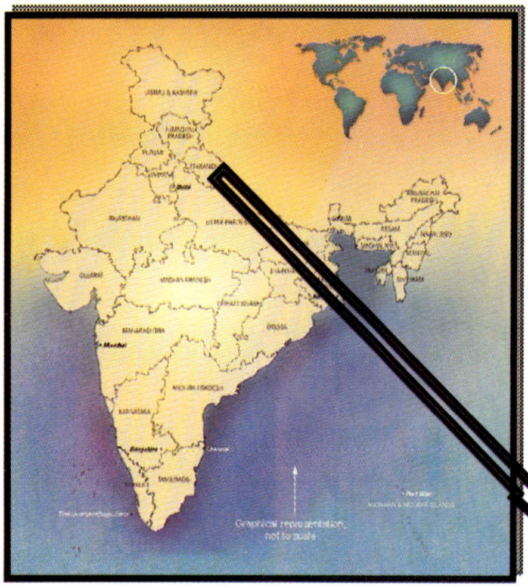

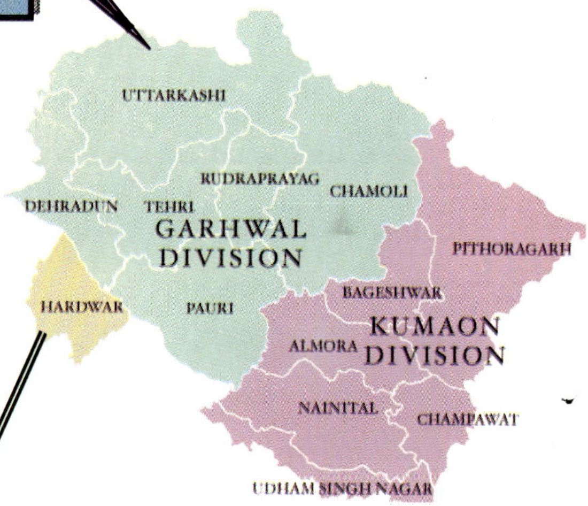

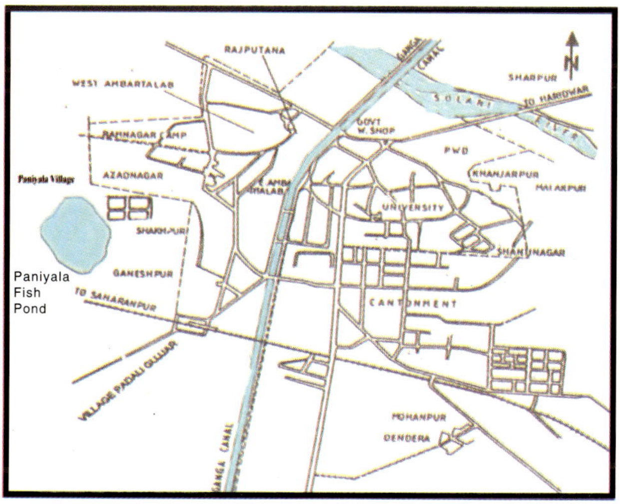

Map 2.1: Showing Paniyala Fish Pond at Roorkee

Figure 2.1: Sampling Site–I at Paniyala Pond

Figure 2.2: Sampling Site–II at Paniyala Pond

Figure 2.3: Sampling Site–III at Paniyala Pond

Figure 2.4: Sampling Site–IV at Paniyala Pond

Chapter 3

Review of Literature

From ancient times people are concerned about the water quality, as they had realized that it was one of the most absolute necessities for life. Ponds are that time also the major source of water used for different purposes from domestic to agricultural. But with the rapid industrialization the condition is now changing. Ponds have now been used to mitigate the increase in runoff rates that is a typical consequence of urbanization. They do this by detaining stormwater runoff for a period of hours or days, while releasing it slowly to receiving streams and lakes. Although designed primarily for control of water-quantity increases, these ponds are now also being used to reduce non-point pollution.

Different workers from India and abroad have discussed various aspects of physico-chemical and biological characteristics of water bodies from time to time. However, it is hard to cite the name of all scientists, limnologists and hydrobiologists who worked on the various aspects of physico-chemical and biological characteristics of water ecology. This review includes some of the important contributions made from various parts of the globe.

Several studies have been done on various water bodies in context of its abiotic and biotic characteristics at national and international level.

Eddy (1934) made a study on freshwater plankters communities. Brune (1953) studied about the water holding efficiency of reservoir. Ruttner (1963), Mackereth (1963) suggested some methods of water analysis. Ruttner (1963) in fundamentals of limnology has excessively deserved limnobiotic components. Limnological aspects of several fresh water ecosystems in India have been studied by Mackereth (1963). Zafar (1966) studied limnological status of Husain Sagar Lake in Hyderabad. Hutchinson (1967), Khatri (1980) and Belsare (1987) studied the hydrobiology of various water bodies. Moitra and Bhowmik (1968) studied the fresh water fisheries pond in Kalyan, West Bengal. Limnological status and fishes of Tungabhadra reservoir was studied by David *et al.* (1969). Shreenivasan (1970) and Murugesan *et al.* (1994) did the comparative study of limnological features of reservoir of Tamil Nadu, Munawar (1970) studied the freshwater pond of Hyderabad, hydrobiology of lower lake of Bhopal regarding to productivity of economic fishes was done by George (1976).

Daborn (1976), Lind (1979) studied the physical and chemical factors of temporary pond in Western Canada. Carlson (1977) and Canfield and Bachmann (1981) presented their study regarding total phosphorus concentration, chlorophyll a and Secchi depth in lakes, Egborge (1978c) have studied the physico-chemical parameters of various freshwater bodies in context of physico-chemical, biological parameters. Egborge (1979) studied the water chemistry of Lake Asejire in Nigeria.

Khan and Zutshi (1980) studied about the primary productivity of Kashmir Himalaya lake, Goel *et al.* (1981) studied the impact of human activities in Hamwa Ramgarh reservoir near Jaipur. Unni (1982) studied the water quality of Sumpna reservoir, phytoplankton of Bhagirathi river and their abundance was carried out by Sharma (1985). Egborge (1986) observed diurnal changes in physico-chemical variable of lake in Nigeria.

Joshi (1987) did the diurnal study of physico-chemical properties of water, study regarding hydrobiological status of Gandhi Sagar reservoir was carried out by Rao *et al.* (1988), Seasonal variation in primary production of upper lake of Bhopal was studied by Gautam (1988), Das (1989) presented his study regarding water quality and water pollution. Khanna (1993) did work on ecology and pollution status of River Ganga. Study of physico-chemical and biological status of industrially polluted Husain Sagar lake was done by Pandey *et al.* (1993). Bachmann (2001), Cooke and Kennedy (2001) presented his study how to manage supply of drinking water.

Many researchers have studied the hydrobiology of the freshwater bodies. Pennak (1955) did the study regarding water cycles for some Colorado reservoir and lakes, Munawar (1970) studied freshwater pond of Hyderabad, Tundisi and Tundisi (1976) worked on the zooplankton ecology of Brca reservoir, Stross *et al.* (1980) studied the zooplankton production, life cycle in pond, Mohanty (1981) studied the water quality of some water bodies of Bhubaneswar. Tirath and Bhargave (1982) studied the water quality in Gomati River at Lucknow. Chattopadhay *et al.* (1984) made a study on the pollution status of river Ganga in Kanpur. Seasonal abundance of phytoplankton in Bhagirathi river was studied by Sharma (1985). Rao *et al.* (1988) respectively gave good account about the morphometry of Gandhi Sagar reservoir and the lower lake of Bhopal.

Various studies have been done in context of quality status by many researchers like Jhingran (1975) reviewed the available broad morphometric features of Indian reservoirs. Limnological studies on the Garhwal Himalayan Rivers were started in 1979 by the work of Badola (1979) who observed the ichthyofauna of some fresh water resources of Garhwal region by an ecological study. Badola and Singh (1981) made a study on fish and fisheries of river Alakananda; they also studied the hydrobiology of the river of Garhwal Himalaya. Dobriyal and Singh (1981) investigated the diurnal variation in some aspect of limnology of river Mandakini from Garhwal Himalaya. Khanna and Badola (1990) observed lchthyofauna of the river Ganga at the foothills of Garhwal Himalaya. Dynamics of limnology with respect to fish production were studied by various other scientists also.

Internationally, limnological studies on temperate water were contributed by Chandler (1944), who studied the western lake Eris for limnological and influence of climatic factors to phytoplankton. Pearsall (1930), Ward and Seibert (1963) studied about the diatoms productivity. Seasonal change in temperature of English lake was studied by Jenkin (1942) while Edmondson (1977) studied about the population dynamics and secondary production *etc.* They laid down firm foundation and provided guidelines and inspiration to the present day workers. Detailed investigations on the relationship between productivity and other ecological parameters were carried out later by Goldman and Wetzel (1963) on the primary productivity of lake County in California, Lund (1972), Macan (1974) studied about the eutrophication of aquatic system. Prati *et al.* (1977) assessed the water quality by single index of pollution.

Limnological aspects of several fresh water ecosystems have been studied by many researchers like Phillips (1927) studied the diurnal alteration in pH of Minnesota lake, Welch (1948), Imvebore (1964) studied fresh water habitat of zooplankton, David *et al.* (1969) on the fisheries of Tungabhadra reservoir, Khan and Zutshi (1970) on primary productivity of Kashmir Himalaya lake, George (1976) on productivity of economic fishes for lower lake of Bhopal, Sharma (1978), Unni (1982) studied the water quality of Sumpna reservoir, Rao *et al.* (1988) studied the hydrobiology of Gandhi Sagar reservoir regarding fisheries, Belsare (1990), Kumar (1995) studied about the pollution of river Mayurakshi in south Bihar, Hutchinson (1967), Monizsette and Marvinic (1978), Sangu and Sharma (1985), Pathak, and Mudgal (2005) studied the fish biodiversity of Virla reservoir in M.P. Viewing the water body's number, ecological information is still insufficient for several reservoirs in India including Madhya Pradesh.

Contribution to the physico-chemical aspects of Indian water bodies has been pointed out by Chandler (1944), Rawson (1960), Sharma (1978) and Biswas (1979) did the comparative study for water quality for different lake. Ruttner (1963), Hussaini (1967) studied the primary production of tropical lake, Herzig (1979) studied the water quality status of lake in Central Europe, Sreenivasan (1970, 1974) did the study of reservoir, Gopal *et al.* (1981) studied the water quality of fresh water reservoir, Dutta *et al.* (1997) studied the physico-chemical characteristics of River Tawi, Kulshreshtha *et al.* (1988), Chapman (1998), Belsare *et al.* (1990) studied about the seasonal change in abiotic factors in upper lake, Kant and Anand (1978) studied the relation between physical factor and phytoplankton in Mansar lake, Badola (1979) did the study on the ecology of fishes in Garhwal region, Goel *et al.* (1981) analyzed the water quality of fresh water reservoir. Gautam (1990) studied the pollution status of mountain river.

The relationship between the physico-chemical parameters and plankters production of pond water and their relation with monthly fluctuation of zooplankters are of great importance and basically very much essential in care of fish culture.

Fishes are more dependent on water temperature, pH, dissolved oxygen, free CO_2, Alkalinity and some other salts for growth and development. Any changes in these parameters may effect the growth, development and maturity of fish. Different causal influences, which determine the quality of water show characteristic change from season to season. Dwivedi *et al.* (2002) studied the physico-chemical factors and algal diversity analysis were carried out on two ponds and found to be moderately polluted and showed a trend of increasing eutrophication. Significant correlations were noted between different parameters. Another attempt was made by Dube *et al.* (2002) on numerous seasonal and perennial ponds, wetlands and swamps which are scattered in southeastern Rajasthan.

Importance of morphometric and hydrobiological details of ponds to assess the nutrient, physico-chemical, biological and fisheries management are being realized in the work of Mortimer (1942), Thresh *et al.* (1944), Govind (1963), Lund (1965), and Palmer (1969). The importance of Indian reservoir has been pointed out by David *et al.* (1969), Sreenivasan (1970 and 1974). Sreenivasan (1970) gave the morphometric features of Stanley reservoir.

In Indian fresh water, plankters information have been given by Chacko and Sreenivasan (1955), Chakraborthy *et al.* (1959), David (1969), Zafar (1966), Rao (1987), Adoni (1985), Rao *et al.* (1988), Kumar (1995), Khanna *et al.* (1999), Suvarna and Sharma *et al.* (2002), Khanna *et al.* (2000). Primary productivity has been considered as a major study by some workers like Sreenivasan (1965), Hussaini (1967), Lewis (1974), Arjariya (2003), Singh and Desai (1980), Khan and Zutshi (1980), Gautam (1988), Jeejabai and Rajendra (1982), Prasad (1956).

Khanna *et al.* (1992) observed seasonal trends in the river Ganga at Sapt-Sarovar, Haridwar. Khanna *et al.* (1993) studied the ecology and seasonal diversity of plankters in spring fed stream of Garhwal Himalaya. Khanna (1993) made a study on ecology and pollution of river Ganga Haridwar. Chopra and Patrick (1994) studied the effect of domestic sewage on self- purification of Ganga water at Rishikesh. Chopra and Rehman (1995) made a study on self purification of physico- chemical properties of Ganga canal water at Jwalapur Haridwar.

Lakshiminarayana *et al.*, (1975) works on prevention of sewage pollution by stabilization of ponds. Ernst *et al.* (1991) works on the toxicity of chlorothalonil to aquatic fauna and the impact of its operational use on a pond ecosystem. Crottereau *et al.* (1999) worked on regulation of bacterial proteolytic activity at natural concentrations in dissolved organic matter in a maritime pond. Gonzalez *et al.* (1999) worked on chemical disturbance of the phytoplankton structure in an artificial pond. Khanna *et al.* (2000) studied on impact of abiotic factors on the phytoplanktonic population in a pond. Dziuban (2002) worked on bacterioplankton and bacteriobenthos of some floodplain lakes in the course of the lower Amur River. Tan *et al.* (2002) worked on main factors influencing the growth of planktonic bacteria in fish pond ecosystem. Dwivedi *et al.* (2003) studied complex dynamics of toxin producing algal species and primary productivity in two water ponds of Faizabad. Khanna and Bhutiani (2003) worked on Sitapur Pond at Haridwar. Burns *et al.* (2004) worked on combined use of cultivation-dependent and cultivation independent methods, which indicates that members of most haloarchaeal groups in an Australian crystallizer pond are cultivable. Douglas *et al.* (2004) studied prehistoric Inuit whalers affected Arctic freshwater ecosystems. Kietlinska *et al.* (2005) worked on nitrogen removal from landfill leachate using a compact constructed wetland and the effect of chemical pretreatment. De *et al.* (2006) worked on the expression of the heat shock protein Hsp 70 in chloride target cells of may fly larvae from motorway retention pond. Derome *et al.* (2006) worked on the parallelism in gene transcription among sympatric lake whitefish ecotypes. Rose *et al.* (2006) worked on pesticide removal from cotton farm tail water by a pilot-scale ponded wetland. Wu and Hahn (2006) worked on the differences in the structure and dynamics of polynucleo-bactos communities in a temperate and subtropical lake, revealed at three phylogenetic levels.

Needham and Needham (1972) gave a guide to study fresh water biology. Crayton and Sommerfield (1979) worked on the composition and abundance of phytoplankters in tributaries of the lower Colorado River. Whitton (1985) studied the ecology of a number of European water bodies. Chopra and Rehman (1995) made a study on self-purification of physico-chemical properties of Ganga canal water at Jwalapur.

Khanna *et al.* (1997) investigated population of green algae in relation to physico-chemical factors of river Ganga. Kaur and Khulbe (1998) studied occurrence, distribution and ecology of water Molds in Nanak Sagar reservoir Singh and Ahmed (1990) worked on comparative study of the phytoplankters of river Ganga and pond of Patna. Khanna *et al.* (1999) studied correlation between abiotic factors and plankters population in river Ganga at Rishikesh (U.P). Khanna and Singh (1999) observed diurnal variation in microbial ecology of Narkewala pond at Ferupur, Haridwar. Khanna *et al.* (1999) studied fishes and their ecology of the river Suswa at Raiwala, Dehradun. Badola and Badola (1999) gave observations on destructive fishing methods and recommendations to protect fish fauna in Garhwal Himalayas. Chugh (2000) studied the seasonal variation in microbial ecology of river Ganga at Haridwar. Khanna *et al.* (2000) observed 20 genera of phytoplankters population of a pond, they revealed that highest phytoplankters population was recorded in winter season while lowest in monsoon season. Khanna *et al.* (2001) studied fluctuation in the population density of Macro-Invertebrates of river Ganga at Pashulok Barrage, Rishikesh. Khanna and Singh (2002) published a book on pond fish ecology and economics. Basu and Mitra (2002) studied on pollution of river Damodar at Burdwan with special reference to metals, COD and pH. Kumar and Sharma (2002) studied the water quality of river Krishni (Part-II Biological characteristics and bio indicators). Kala and Sharma (2002) worked on effect of physico-chemical factors on phytoplankters population.

Mruthunjaya and Hosmani (2002) worked on comparative account of physico-chemical parameters, enzymatic activities and plankters population in three water bodies of Mysore city. Naz *et al.* (2002) studied aquatic fungi as an indicator organism Monichan studied on distribution status of fish fauna in Etawah district U.P. Khanna and Bhutiani (2003) worked on limnological status of Satikund pond at Haridwar and they observed that phytoplankters concentration was dominated by diatoms and recorded 3 genera of zooplankters. Shaikh and Yeragi (2003) studied the biodiversity of plankters in Tansa river of Thane district. Prasad and Singh (2003) studied the composition, abundance and distribution of phytoplankters and zoo benthos in a tropical water body. Arjariya (2003) studied the physico-chemical profile and plankters diversity of Ranital lake. Jena *et al.* (2003) studied on water quality parameters around Sagar Island. Maruthanayagam *et al.* (2003) studied the qualitative analysis of zooplankters population of pond indicators and reported occurrence of 21 species during both summer and rainy season. Khanna and Bhutiani (2004) worked on fish and their ecology of river Ganga at Gohri Ghat, Garhwal. Negi and Malik (2006) studied Piscine diversity of district Kumaon.

Oertli *et al.* (2000) added a criterion linked more to water-body function, defining a pond as: 'A water body with a maximum depth of not more than 8 m, offering water plants the potential to colonise almost the entire area of the pond'. At a regional level, ponds can contribute highly to freshwater biodiversity, with recent evidence showing that they often support considerably more species, more unique species and more scarce species than other water-body types (Williams *et al.*, 2004). Moreover, a review of the habitat preferences of a selection of aquatic fauna and flora (Oertli *et al.*, 2000) demonstrated that most species are potential inhabitants of ponds, with a significant proportion of them only ever found in ponds. Ponds may support more species than rivers (*e.g.* invertebrates and plants), and more uncommon species. In terms of regional (gamma) diversity, they often make a greater contribution than any other aquatic habitat (Biggs *et al.*, 2005). However, a crucial issue is to estimate accurately the biodiversity hosted by individual ponds. Many contributions to this special issue highlight relationships between the species richness of various taxa and environmental variables (*e.g.* elevation, hydrology, pond area, water chemistry, habitat heterogeneity) with differences between permanent and temporary ponds (Della Bella *et al.*, 2005; Solimini *et al.*, 2005), and between lowland and alpine ponds (Hinden *et al.*, 2005). Ponds, and the biodiversity they host, are highly sensitive to land use and/or human activities in surrounding areas (Carchini *et al.*, 2005). Disturbance such as to nutrient loads (*e.g.* in farmland or pasture landscapes) may rapidly cause shifts towards eutrophic or turbid conditions (Ruggiero *et al.*, 2005), and affect the species richness of many taxa (Menetrey *et al.*, 2005).

In recent years there has been an increased interest in small lentic (still water) water systems (Oertli *et al.*, 2009). Reasons for this include their significant contribution to regional diversity (Williams *et al.*, 2004, Davies *et al.*, 2008, Ruggiero *et al.*, 2008), including the presence of rare species (Scheffer *et al.*, 2006), and their use as model systems (De Meester *et al.*, 2005). Interest has also risen because pond environments are often neglected habitats (De Meester *et al.*, 2005) or are filled in by natural or artificial processes (Williams *et al.*, 1998).

Ponds are defined as standing water bodies up to 2ha in size holding water for at least 4 months of the year (Biggs *et al.*, 1998). To date the majority of studies have focussed upon ponds within rural and agricultural landscapes. By contrast, ponds within urban areas are poorly understood, with only a few studies to date which begin to explore the links between their biodiversity and environment, *e.g.* Birch and McCaskie, 1999; Stoianov *et al.*, 2000; Wood and Barker, 2000; Gledhill *et al.*, 2008; Vermonden *et al.*, 2009.

However, the water of the ponds, lakes and river are polluted mainly due to discharged waste water from residential areas, sewage outlets, solid wastes, detergents, automobile oil wastes, fishing facilities and agricultural pesticides from farmlands (Srivastava *et al.*, 2003; Usha *et al.*, 2006; Hasan *et al.*, 2007). In recent years, their importance has somewhat declined due to technological advancements leading to more centralized water supply systems. There is a relationship among ecologists and microplanners about the importance of conservation of ponds as sustainable source of water for rural communities (Park and Park, 2005). The present study is an attempt to assess the water quality of ponds of State Fish Department situated at Paniyal Village, Roorkee Distt. Haridwar. The water available may be sustainably exploited for multiple uses like rural water supply, fisheries and even recreation.

Chapter 4
Materials and Methods

Any change in the quality of water has been determined by analysis and ensuring the qualitative and quantitative change in the parameters (physico-chemical and biological parameters) of the water. In this thesis we used standard methods to analyze the different parameters studied.

For present study of the Paniyala Pond, the water samples were collected monthly from different sampling sites (I, II, III, and IV) during March 2007 to February 2009 in morning hours (from 7.00 A.M. to 10.00 A.M.). The samples were taken in borosil glass bottles of 300ml, plastic cans of 1.00 L from each location.

The collection of water samples and analysis was done using standard methods with the help of Welch (1948), Trivedi and Goel (1984), APHA-AWWA-WPCF (1998) and Khanna and Bhutiani (2004) for physico-chemical parameters. Biological parameters were analysed with the help of Jhingran (1975), Badola (1979), Khanna (1993) and APHA-AWWA-WPCF (1998).

Sampling Programme and Analysis

Bearing in mind the pond characteristics and the water quality requirements, the following routine sampling and analysis programmes were devised:

Number of sites	:	Four
Collection of samples	:	Monthly
Time of sample collection	:	Between 7:00 A.M. to 11:00 A.M.
On site analysis	:	Temperature and pH
Analysis within 24 hours of sampling	:	All physico-chemical, biological parameters and microbiological analysis taken under study

Chemicals

The chemicals used in the present investigation were of analytical grade and were supplied by E. Merck India, S.D. fine chemicals and B.D.H., India.

Instruments

Following instruments were used during the study period.

Sl.No.	*Name of Instruments*	*Make/ Manufactured by*
1.	BOD incubator	Tasco
2.	Electronic balance	Roy Electronic
3.	Jackson turbidity meter	–
4.	COD Assembly	–
5.	Conductivity Meter	Microprocessor based Century
6.	Hot plate	Sunvic
7.	Hot air oven	Modern Scientifics
8.	Microscope	–
9.	Plankton collection net	–
10.	Sedgwick rafter counting cell	–
11.	Colony counter	–
12.	Laminar air flow	–
13.	Deep freezer	–
14.	Mantle heater	A.J. Enterprises
15.	pH meter	Tasco
16.	Rotatory flask shaker	Tanco
17.	Refrigerator	Godrej
18.	Thermometer (°C)	–
19.	Distil water plant	Tanco
20.	Glass fiber filter	–
21.	Dessicator	–
22.	Autoclave	–

Sampling Methods

(A) Cleaning and Sterilization of Glassware

All the glasswares and jerricans were cleaned or washed with soda and chromic acid and rinsed with double distilled water 2-3 times.

(*i*) Sample Bottles

300 ml capacity BOD bottles made of borosil were used for sample collection. They were washed with washing soda and chromic acid and rinsed with tap water followed by distilled water; the neck and stopper were wrapped with butter paper with the help of rubber band. The bottles were then sterilized in an autoclave at 15-lbs./inch^2 pressure (121°C) for 15-25 minutes.

(*ii*) Pipettes

Pipettes of different volume size were washed and fitted with cotton plug at the upper end; these were then wrapped in butter paper and sterilized in an autoclave at 15-lbs./inch^2 pressure at 121°C for 15-25 minutes.

(*iii*) Test Tubes

Borosil test tubes were washed and then plugged with non-absorbent cotton wool, these were arranged in test tube racks and sterilization was done in an autoclave at 15-lbs./inch^2 pressure at 121°C for 15-25 minutes.

(*iv*) Other Glasswares and Materials

Such as conical flasks, beakers, measuring cylinders, petri-plates, inoculation loop, fermentation tubes, Durham's tube, culture tube, biological pipettes, spirit lamp, cotton, filter paper, water bath *etc.* were washed carefully and then sterilized in hot air oven at 160-180 °C for 2-3 hours.

(B) Collection of Samples

(*i*) For Determination of the Parameters

For the estimation of total solids, TDS, TSS, turbidity, COD, alkalinity, pH (hydrogen ion concentration), hardness and chlorides samples were taken from 2 feet below the surface water by keeping and opening jerricans below the surface water. Caps of cans were removed after dipping the can and also closed in the water after filling up of can.

(*ii*) For Dissolved Oxygen/Biochemical Oxygen Demand

Samples were collected in sterilized 300 ml capacity BOD bottles. The bottles were filled completely with sample water up to the rim and stopper was placed inside water only avoiding any kind of air bubble inside it. DO was immediately fixed by adding 2 ml each of manganous sulphate ($MnSO_4$) and alkaline potassium iodide at sampling spot, while the sample for BOD was incubated for 5 days in BOD incubator at 20 °C.

(*iii*) Sampling for Biological Analysis

For biological analysis of water, samples were collected with the help of plankton net attached with a collection tube at the base of net. For this a known amount of water was filtered through the plankton collection net and sample was collected inside collection tube. The sample was then transferred in sterilized tube of 50 ml capacity and preserved by adding one drop of lugol's solution or formaldehyde solution.

(*iv*) Sampling for Microbiological Analysis

For microbiological analysis collection was made in sterile bottle to avoid any kind of contamination. The sterilized sampling bottle wrapped in butter paper opened at sampling site and after collecting the sample were closed immediately. These bottles were opened in sterile conditions.

(C) Reagent Preparation

(*i*) Potassium Dichromate Solution (0.025N)

Dissolve 12.259 gm of dried A.R. grade $K_2Cr_2O_7$ in distilled water to make one liter of solution. Dilute it 10 times to make 0.025N $K_2Cr_2O_7$ solution.

(*ii*) Ferrous Ammonium Sulphate (0.01N)

Dissolve 39.2 gm of Fe $(NH_4)_2(SO_4)_2.6H_2O$ in distilled water, adding 20 ml of concentrated H_2SO_4 to make 1 liter of solution. Dilute it 10 times to make 0.01 N Fe $(NH_4)_2(SO_4)_2.6H_2O$ solution.

(*iii*) Ferroin Indicator

Dissolve 1.485 gm of 1, 10- Phenonthroline and 0.695 gm of ferrous sulphate ($FeSO_4.7H_2O$) in distilled water to make 100 ml of solution.

(*iv*) Sulfuric Acid

H_2SO_4 Concentrated (Sp. Gr.1.84)

(*v*) Mercuric Sulphate

$HgSO_4$ Solid

(*vi*) Sodium Thiosulphate (0.025N)

Dissolve 24.82 gm of $Na_2S_2O_3.5H_2O$ in boiled distilled water to make final volume to one liter. A pellet of NaOH or 0.4 gm borax was added as stabilizer. This is 0.1 stock N solution. Dilute 0.1N stock solution to 4 times with boiled distilled water to prepare 0.025N solution. Keep it in brown glass stopper bottle.

(*vii*) Alkali Iodide Azide Solution

(*a*) Dissolve 500 gm of NaOH or 700 gm of KOH and 150 gm of KI in distilled water to makes one liter of solution.

(*b*) Dissolve 10 gm of NaN_2 in 40 ml of distilled water mix both (a) and (b) solution.

(*viii*) Manganous Sulphate Solution

Dissolve 100 gm of $MnSO_4.4H_2O$ in 200 ml of boiled distilled water and filter.

(*ix*) Starch Solution

Dissolve 1 gm of starch in 100 ml of warm distilled water and add a few drops of formaldehyde solution.

(*x*) Sodium Hydroxide (0.05N)

Dissolve 40 gm of NaOH in one liter boiled CO_2 free distilled water to prepare 1N NaOH stock solution. Dilute 50 ml of 1N NaOH stock solution to 1 liter distilled water. Standardized the solution using H_2SO_4.

(*xi*) Phenolphthalein Indicator

Dissolve 0.5 gm of phenolphthalein in 50ml of 95 per cent ethanol and add 50 ml of distilled water. Add slowly 0.05N CO_2 free NaOH until solution turns faintly pink.

(*xii*) Methyl Orange Indicator

Dissolve 0.5 gm methyl orange in 100 ml of distilled water.

(*xiii*) Hydrochloric Acid (0.1N)

Dilute 12N concentrate HCl (sp.gr.1.18) to 12 times (8.34—100ml) to prepare 1.0N HCl. It is further diluted ten times to make 0.1N HCl (100—1000ml).

(*xiv*) EDTA Solution (0.01M)

Dissolve 3.723 gm of disodium salt of EDTA in distilled water to prepare one liter of solution.

(*xv*) Buffer Solution

(*a*) Dissolve 16.9 gm of ammonium chloride (NH_4Cl) in 143 ml of concentrated ammonium hydroxide (NH_4OH).

(*b*) Dissolve 1.179 gm of disodium EDTA and 0.78 gm of $MgSO_4.7H_2O$ in 50 ml distilled water.

Mix (*a*) and (*b*) solution and make final volume to 250 ml using distilled water.

(*xvi*) Eriochrome Black-T Indicator

Mix 0.5 gm of EBT with 100 gm of NaCl (A.R) and grind.

(*xvii*) Sodium Hydroxide Solution (1N)

Dissolve 40 gm of NaOH in one liter distilled water.

(*xviii*) Silver Nitrate, (0.02N)

Dissolve 3.40 gm of dried $AgNO_3$ (A.R) in distilled water to make one liter of solution and kept in dark bottle.

(*xix*) Potassium Chromate (5 per cent)

Dissolve 5 gm of K_2CrO_4 in 100 ml of distilled water.

(*xx*) Ethyl Alcohol, 95 per cent

(*xxi*) Lugol's Solution

Dissolve 10gm of KI and 5 gm of iodine crystals in 20 ml of distilled water and then add 50 ml distilled water in which 5 gm anhydrous sodium acetate has been dissolved.

(D) Analytical Methods

Samples were analyzed for the following parameters:

(i) Physico-chemical Parameters

(*a*) Water temperature

(*b*) Conductivity

(*c*) Turbidity
(*d*) T. S.
(*e*) pH
(*f*) DO
(*g*) COD
(*h*) BOD
(*i*) Free CO_2
(*j*) Total alkalinity
(*k*) Hardness Ca^{++} / Mg^{++}
(*l*) Chlorides

(*ii*) Biological Parameters
(*a*) Plankters
(1) Zooplankters
(2) Phytoplankters
(*b*) Aquatic Flora and Fishes

(*iii*) Microbiological
(*a*) MPN
(*b*) SPC
(*c*) Aquatic fungi

Methodology

The collection of water samples and analysis was done with the help of Welch (1948), Trivedy and Goel (1984), APHA-AWWA-WPCF (1998) and Khanna and Bhutiani (2005) for physico-chemical parameters. Biological parameters were analysed with the help of Jhingran (1975), Badola (1979), Khanna (1993) and APHA-AWWA-WPCF (1998).

(a) Temperature

Centigrade calibrated thermometers were used to record the temperature of surface water for measuring water temperature. Thermometer was dipped into water up to a desirable depth for 5-7 minutes and reading was taken.

(*b*) Total Solids

Solids refer to matter suspended or dissolved in water or wastewater. Solids may affect water or effluent quality adversely in a number of ways. Total Solid is a term applied to the material residue left in a vessel after evaporation of a sample and its subsequent drying in an oven at a defined temperature.

Procedure

A 50ml washed and dried beaker was taken and weighted. 20ml of unfiltered sample was taken in that beaker and sample was evaporated over hot plate/water bath and then dried in an oven at 105 ±1°C for one hour. The final weight of the beaker was taken after drying of sample.

Calculation

$$\text{Total Solids (mg/l)} = \frac{(A - B) \times 1000 \times 1000}{V}$$

where,

A: Final weight of beaker in gm

B: Initial weight of beaker in gm

V: Volume of sample taken in ml

(c) Conductivity

The conductivity of water was directly measured by the help of conductivity meter and was recorded as µ mhos/cm.

(d) Turbidity

Clarity of water is important in producing product destined for human consumption and in many manufacturing operations. The clarity of a natural body of water is an important determinant of its condition and productivity.

Turbidity in water is caused by suspended and colloidal matter such as clay, silts, finely divided organic and inorganic matter, plankton and other microscopic organisms. Turbidity is an expression of the optical property that causes light to be scattered and absorbed rather than transmitted with no change in direction or flux level through the sample. The turbidity of water was measured with the help of Jackson's Candle Turbidity Meter. It consists of three parts (a) Candle (b) Holder (c) Calibrated glass tube

Procedure

Collected samples were shaked thoroughly in jerricans and were slowly transferred into calibrated glass tube of Jackson's Candle Turbidity Meter until the flame of the candle disappears and the light path was noted in centimeters (measured from inside the bottom of the glass tube). The reading (cm) was then converted in to Jackson Turbidity Unit (JTU) using conversion table.

(e) pH

Measurement of pH is one of the most important and frequently used tests in water chemistry. pH is the hydrogen ion activity or concentration in the given water sample. pH equal to negatively $\log_{10}$ of hydrogen ion concentration

$$pH = -\log_{10}(H^+)$$

pH as defined by Sorenson (1909) is $-\log_{10}(H^+)$; it is the "intensity" factor of acidity. pH of water sample was measured by the use of pH strips, but the accurate method is the electrometric method. The sample was analyzed with the help of electronic digital pH meter consisting of a reference and indicator electrode.

Procedure

Calibrate the electrode system against standard buffer solution of known pH. Rinse the electrode using distilled water, now dip the electrode in sample water and note the reading on digital screen.

(*f*) Dissolved Oxygen (DO)

Dissolved oxygen (DO) level in natural water and wastewater depend on the physical, chemical and biochemical activities in the water body. The analysis for DO is a key test in water pollution and waste treatment process control.

DO is the measure of oxygen concentration present in a given water sample. The concentration of oxygen reflects whether the process undergoing is aerobic or anaerobic.

Principle

The manganous sulphate reacts with the alkali (KOH, NaOH) to form a white precipitate of manganous hydroxide, which in the presence of oxygen, gets oxidized to a brown colour compound. In the strong acid medium manganic ions are reduced by iodide ions which get converted into iodine equivalent to the original concentration of oxygen in the sample. The iodine can be titrated against thiosulphate using starch as an indicator.

Certain oxidizing and reducing materials may effectively interfere with the determination of oxygen by converting iodide ions to iodine or vice-versa. The azide modification removes the interference substance specially nitrite, which is destroyed by sodium azide (NaN_2). That's why azide modification method is used to determine dissolved oxygen instead of Winkler's iodometric method.

Procedure

Fill sample water in 300 ml capacity BOD bottle up to the rim and stopper is placed inside water body avoiding any kind of air bubble inside it. Add 2 ml each of $MnSO_4$ and alkaline iodide azide solution using separate pipettes. A precipitate was appeared. The bottle was kept for some time after mixing thoroughly. At this stage the DO was fixed. 2 ml of Concentrated H_2SO_4 was added to dissolve the precipitate by shaking it. The part or whole of the content was titrated against sodium thio-sulphate using starch as an indicator. At end point blue colour changes to colourless.

Calculation

When whole contents have been titrated

$$\text{Dissolved Oxygen (mg/l)} = \frac{(\text{ml} \times \text{N}) \text{ of titrant} \times 1000 \times 8}{V_1 - V}$$

When only a part of content has been titrated

$$\text{Dissolved Oxygen (mg/l)} = \frac{(\text{ml} \times \text{N}) \text{ of titrant} \times 1000 \times 8}{\dfrac{V_2(V_1 - V)}{V_1}}$$

where,

V_1: Volume of sample bottle after placing stopper
V_2: Volume of part of the contents titrated
V: Volume of $MnSO_4$ and KI added.

(*g*) Biochemical Oxygen Demand (BOD)

The biochemical oxygen demand (BOD) determination is an empirical test in which standardized laboratory procedures are used to determine the relative oxygen requirement of wastewater effluents, and polluted waters.

The test measure the molecular oxygen utilized during a specified incubation period for the biochemical degradation of organic matter (carbonaceous demand) and the oxygen used to oxidize inorganic material such as sulfides and ferrous ions. In other words biological oxygen demand, (BOD) represents the amount of oxygen required for the microbial decomposition of the organic matter in the water. The BOD procedure, which is used extensively in monitoring water quality and biodegradation of waste material, is designed to determine how much oxygen microorganisms consume during oxidation of the organic matter in the sample.

Principle

The principle of the method involves, measuring the difference of the oxygen concentration between the samples after incubating it for 5 days at 20°C.

Procedure

Samples were collected in two BOD bottles of 300 ml capacity from each site. One set of the bottle were kept in BOD incubator at 20°C for 5 days and the DO constant was determined in another set immediately. After the compilation of 5 days incubation, the DO of first set was determined and BOD was calculated by means of given formula:

Calculation

$$\text{BOD (mg/l)} = D_0 - D_5$$

where,

D_0: Initial DO in the sample
D_5: DO after five days incubation

(*h*) Chemical Oxygen Demand (COD)

Chemical oxygen demand (COD) is defined as the amount of specified oxidant that reacts with the sample under controlled conditions. COD is widely used to characterize the organic strength of wastewater and pollution of natural water. It is the measure of oxygen consumed during the oxidation of the oxidizable organic matter by a strong oxidizing agent.

Principle

The sample is refluxed with $K_2Cr_2O_7$ and H_2SO_4 in presence of mercuric sulphate, to neutralize the effect of chlorides and silver sulphate, which acts as a catalyst. Excess potassium dichromate ($K_2Cr_2O_7$) is titrated against ferrous ammonium sulphate using ferroin as an indicator. The amount of $K_2Cr_2O_7$ used is proportional to the oxidizable organic matter present in the sample.

Procedure

20 ml of sample was taken in a 250 ml capacity COD flask; slowly add 10 ml of $K_2Cr_2O_7$ and a pinch of $HgSO_4$ and Ag_2SO_4 were added. Add slowly 30 ml of concentrated H_2SO_4 and reflux condenser was connected to the COD flask, reflux for 2 hours on hot plate. Now remove the flask, cooled and make the final volume to 140 ml in flask using distilled water, take suitable amount of sample in conical flask add 2-3 drops of ferroin indicator and titrate against 0.01N ferrous ammonium sulphate.

Run a blank with distilled water using same quantity of chemicals.

Calculation

$$\text{COD (mg/l)} = \frac{(\text{B} - \text{A}) \times \text{N of Ferrous ammonium sulphate} \times 1000 \times 8}{\text{ml of sample}}$$

where,

A: ml of titrant with sample

B: ml of titrant with blank

(*i*) Free CO_2

Ten drops of neutralized phenolphthalein indicator was added to water sample (known amount) and titrated by N/44 NaOH until the pink color flashed throughout the solution. The volume of titrant used was recorded and estimation of CO_2 was made by the following formula:

$$\text{Free CO}_2\text{ (mg/l)} = \frac{(\text{ml} \times \text{N of titrant}) \times 1000 \times 44}{\text{ml of sample taken}}$$

(*j*) Total Alkalinity

Alkalinity of water is its acid- neutralizing capacity. It is the sum of all the titrable bases.

Principle

Total alkalinity, carbonates and bicarbonates can be estimated by titrating the sample with a strong acid (HCl or H_2SO_4), first to pH 8.3 using phenolphthalein as an indicator and then further to pH between 4.2 and 5.4 with methyl orange.

Procedure

Take100 ml of sample in a conical flask and add 2 drops of phenolphthalein indicator, if solution remain colourless then phenolphthalein alkalinity (PA) =0 *i.e.* absent and total alkalinity was determined, if solution turns pink, titrate it against 0.1 N HCl until color disappears this is phenolphthalein alkalinity (PA). Add 2-3 drops of methyl orange indicator to the same sample and titrate against 0.1N HCl until yellow color changes to pink at the end point this is total alkalinity (TA).

Calculation

$$\text{Alkalinity (mg/l)} = \frac{(\text{ml} \times \text{N}) \text{ of HCl} \times 1000 \times 50}{\text{Volume of sample}}$$

(*k*) Hardness Ca^{++}/ Mg^{++}

Water hardness was understood to be a measure of the capacity of water to precipitate soap. Soap is precipitated chiefly by the calcium and magnesium ions present.

Principle

If a small amount of dye such as Eriochrome Black T [EBT] is added to an aqueous solution containing calcium and magnesium ions at a pH of 10.0 ±0.1, the solution becomes wine red. If EDTA is added as a titrant, the calcium and magnesium will be complexed and solution turn from wine red to blue, marking the end point of titration.

Procedure

Take 100 ml of sample in a conical flask and add a pinch of EBT indicator and one ml of buffer solution, the solution turns wine red, titrate it against EDTA until wine red color change to blue.

Calculation

$$\text{Hardness (mg/l)} = \frac{\text{ml of EDTA} \times 1000}{\text{ml of sample}}$$

(i) Ca^{++} Hardness

100 ml of sample was taken in a flask, followed by the addition of murexide indicator and 11ml of 1N NaOH. After it the sample was titrated by standard EDTA solution till the color changed (pink colour change to purple) this is end point. This is calculated by understated equation:

Calculation

$$Ca^{++} \text{ Hardness (mg/l)} = \frac{\text{ml of EDTA} \times 1000}{\text{ml of sample titrated}}$$

(ii) Mg^{++} Hardness

Magnesium was estimated by subtracting the calcium hardness from total hardness. It was calculated by using formula:

Mg^{++} hardness (mg/l) = Total hardness – Ca^{++} hardness.

(*l*) Chloride

The chloride concentration is higher in wastewater than in raw water because sodium chloride (NaCl) is a common article of diet and passes unchanged through the digestive system.

Principle

Silver nitrate reacts with chloride to form very slightly soluble white precipitate of AgCl. At the end point when all the chlorides get precipitated, free silver ions react with chromate to form silver chromate of reddish brown colour.

Procedure

Take 50 ml of sample in a conical flask and add 2 ml of potassium chromate to it, titrate the solution against silver nitrate until persistent red tinge appears.

Calculation

$$\text{Chloride (mg/l)} = \frac{(\text{ml} \times \text{N}) \text{ of } AgNO_3 \times 1000 \times 35.5}{\text{ml of sample}}$$

Biological Examination

Plankton

The term 'plankton' refers to those microscopic aquatic forms having little or no resistance to currents and living free floating and suspended in natural water. The phytoplankton (microscopic algae) occurs as unicellular, colonial, or filamantous forms. The zooplankton in fresh water comprises principally protozoan, rotifers, cladocerans and copepods.

Collection of Plankton

The plankton of pond water was collected by fine plankton collection net. A known amount of water was filtered through plankton collection net and concentrated in desired quantity inside collection tube. The sample was preserved in Lugol's solution.

Procedure

1 ml of concentrate was taken and placed in Sedgwick rafter counting cell, a glass slip was placed over it avoiding any kind of bubble. Calculation of the organisms was done by applying the following formula: -

Calculation

$$\text{Plankton (unit/l)} = \frac{(a \times 1000)\, C}{L}$$

where,

a: Average number of plankton in one small counting chamber of Sedgwick rafter counting cell

C: ml. of plankton concentrate

L: Volume of original water filtered in liter.

Fish Diversity

Fishes were collected by using jal, fatiyala (triangular net) and fine clothes. Fishes were preserved in 5 per cent formaline and identified with the help of Badola (1979) and Jhingran (1975).

Microbiological Parameters

(*i*) Most Probable Number (MPN)

The bacteriological quality of water is based on testing for bacterial species of known excretal origin, particularly organisms of the coliform group. The coliform group of bacteria included all the aerobic and facultatively anaerobic, Gram negative, non-sporulating bacilli, that produce acid and gas from the fermentation of lactose.

Coliform bacteria as typified by *E. coli* and faecal *Streptococci* residing in the intestinal tract of humans are excreted in faeces of humans and other warm-blooded animals averaging about 50 million coliforms per gram.

The multiple tube fermentation technique can be used to enumerate positive presumptive and confirmed coliform tests. Results of the test are expressed as the Most Probable Number (MPN), since the count is based on statistical analysis of sets of tube in a series of serial dilution. MPN is by definition is related to a sample volume of 100 ml.

Apparatus and Material

1. Fermentation tubes (50 ml)

Presumptive Test

MacConkey's Broth is required for this test. The composition of broth is as given below:

Bile salt or sodium taurocholate	-	5.00 gm
Peptone	-	20.00 gm
Lactose	-	10.00 gm
NaCl	-	5.00 gm
1 per cent Bromo cresol purple	-	1.00 ml
Distilled water	-	1.00 litre
Final pH (after sterilization)	-	7.10±0.20

Procedure

Several dilutions of the sample for river water were prepared. Dilutions above than 1:1000 are required. 5 fermentation tubes for each dilution were filled with 9 ml of broth and one Durham's vial was put in each tube in inverted condition. Now all the tubes were plugged with cotton and sterilized in autoclave. All the dilutions were shaked vigorously before moving the sample to fermentation tubes. Separate sterilized pipettes were used for different dilutions as well as samples. Fermentation tubes were gently shaked. Now these tubes were incubated at 35-37 °C temperature for 48 hours. All the tubes were examined after 48 hours for gas production.

The tube showing gas production in Durham's tube, were recorded as positive (+ve). Even a tiny bubble should be recorded as positive. The tubes showing positive results were now subjected to the confirmatory test. Negative tubes were incubated for further 24 hours to confirm the test.

(*i*) Confirmatory Test

For this test Brilliant Green Lactose Bile (BGLB) Broth is required. This inhibits the growth of lactose fermenters other than coliform; thus, gas formation in BGLB broth constitutes a confirmed test *i.e.* coliform present. The composition of BGLB broth is given below.

Peptone	-	10.0g
Lactose	-	10.0g
Oxgall	-	20.0g
Brilliant green	-	0.0133g
Distilled water	-	1.0l
pH (after sterilization)	-	7.2 ±0.2

Procedure

Fermentation tubes with 10ml BGLB broth and inverted Durham's tube were prepared. The number of tubes should be equal to all positive tube in the presumptive test. Positive tubes were shook gently. One loopful of liquid was transferred to the tubes having BGLB with the help of inoculation loop. These tubes were incubated at 35–37°C for 48 hours. The tubes were recorded as positive, which showed the gas production. The number of positive tubes were put in MPN table and then in the formula to calculate the MPN/100 ml.

Calculation

$$\text{MPN}/100\ \text{ml} = \frac{\text{MPN table value} \times 10}{\text{Starting value}}$$

(*ii*) Standard Plate Count (SPC)

On a solid medium, counting of organism depends upon the fact that living cells will produce sufficient progeny to form a colony visible to naked eyes. Since bacteria occurring in water are single cell, pairs, groups or even dense clumps, not every individual living cell will develop into a separate colony on incubation. Therefore a number of colonies appearing on plate does not necessarily represent the total number of organisms present in the test volume. The results are expressed as the cell forming units (CFU) per ml.

Materials

1. Nutrient Agar Medium (NAM)

Nutrient Agar Medium (NAM)

Beef extract	-	3.0 g
Peptone	-	5.0 g
Agar agar	-	15.0 g
Distilled water	-	1.0 L
pH (after sterilization)	-	6.8 ± 0.2

Preparation of Serial Dilution

5 or 6 culture tubes each containing 9ml distilled water, were sterilized after plugging. The sample was vigorously shaken and 1 ml was transferred to the first

culture tube. Now 1 ml water from first tube was again transferred to 2nd test tube and so on. Separate pipettes were used for each transfer. The dilutions were 1:10, 1:100,1:1000, 1:10000 and so on.

Procedure

3 pairs of borosil petridishes were sterilized in an oven. Now 1 ml of liquid from last 3 dilution was transferred to the petridishes by opening them gently. 15-20 ml medium was poured into these plates. The temperature of NAM should be about 45°C. Both water sample and medium, were mixed by gently rotating the plate. When medium became solidify the plates were inverted and incubated at 37± 2°C temperatures for 48 hours. The colonies were counted after 48 hours by colony counter and Colony Forming Unit (CFU)/ml was calculated.

Calculation

$$\text{CFU/ml} = \frac{\text{Colonies counted}}{\text{Dilution factor}}$$

Aquatic Fungi

Aquatic fungi are found wherever non living organic matter occurs, although some species are pathogenic and other are parasitic, unpolluted stream water has relatively large number of species representing the aquatic fungi (species possessing flagellated zoospores and gametes) aquatic hypomycetes and soil fungi.

Procedure

1 ml of sample from selected dilutions were transferred into sterilized Petri plates by opening them gently. Now 15-20 ml of Rose Bengal media was slowly poured in these plates the temperature of media should be about 45°C. Sample and media were mixed gently by rotating the plates. When medium solidifies plates were incubated at 27°C ± 2 temperatures for 48 hours in an inverted position. The colonies were counted after 2-3 days by colony counter and colony forming unit (CFU)/ml. was calculated.

$$\text{CFU/ml} = \frac{\text{Colonies counted}}{\text{Dilution factor}}$$

Neo-peptone- Glucose Rose Bengal Aureomycin Agar (NPGRBAA) is required for this test. The composition of broth is as given below.

Neopeptone	-	5.0g
Glucose	-	20.0g
Rose Bengal solution	-	10.0g
Agar	-	5.0g
Distilled water	-	1.0L
Final pH (after sterilization).	-	6.5+0.2

Prepare Rose Bengal solution by dissolving 1 gm Rose Bengal dye in 100 ml distilled water.

Chapter 5

Parameters for Assessing Pond Water Quality

PHYSICAL CHARACTERISTICS

Sampling Site–I

The physical characteristics of sampling site–I are presented in Table 5.I.1 and Figures 5.I.1 and 5.I.2.

(*i*) Water Temperature

The water temperature at this sampling site was observed minimum of 17.1 °C in the month of December for year 2007-2008 and maximum of 30.1 °C in June in the year 2007-2008. The average value in the year 2007-2008 was found to be 25.56 ° C ± 4.51 and 25.52 ° C ± 4.27 in the year 2008-2009.

(*ii*) Conductivity

The conductivity of pond water was observed to be minimum 28.00 μmhos/cm^2 in the March month in the year 2007-2008 and maximum of 98.00 μmhos/cm^2 in the September Month in the year 2007-2008. The average values recorded in the year 2007-2008 was 61.08 μmhos/cm^2 ± 26.43 and 62.50 μmhos/cm^2 ± 23.23 in the year 2008-2009.

(*iii*) Turbidity

The turbidity in water was minimum 17.00 JTU in the month of November in the year 2007-2008 and maximum of 104.00 JTU in the month of July in the year 2007-2008. The average value of Turbidity in the year 2007-2008 was 52.50 JTU ± 33.06 and 53.83 JTU ± 32.49 in the year 2008-2009.

(iv) Total Solids

The total solids were minimum 52.15 mg/l in November month in the year 2008-2009 and maximum of 779.23 mg/l in the July month in the year 2008-2009. The average value of total solids in the year 2007-2008 was 296.18 mg/l± 236.89 and 308.53 mg/l ± 250.33 in the year 2008-2009.

Table 5.I.1: Monthly Variation in Physical Parameters of Sampling Site–I, 2007-2009

Months	*Water Temperature (°C)*	*Conductivity (μmhos/cm²)*	*Turbidity (JTU)*	*Total Solids (mg/l)*
		2007-2008 (Sampling Station – I)		
March	25.90	28.00	41.00	207.05
April	27.70	30.00	69.00	254.05
May	28.90	39.00	74.00	243.42
June	30.10	39.00	92.00	453.50
July	29.80	42.00	104.00	720.54
August	27.20	50.00	98.00	642.54
September	28.40	98.00	42.00	564.21
October	29.30	97.00	18.00	126.41
November	21.20	93.00	17.00	67.41
December	17.10	75.00	20.00	61.56
January	19.20	78.00	26.00	98.01
February	21.90	64.00	29.00	115.43
Average ±Sd	25.56±4.51	61.08±26.43	52.50±33.06	296.18±236.89
		2008-2009 (Sampling Station – I)		
March	25.20	42.00	39.00	251.01
April	27.90	36.00	72.00	262.45
May	29.20	40.00	79.00	243.15
June	30.00	35.00	90.00	412.23
July	29.40	43.00	102.00	779.23
August	26.80	51.00	99.00	678.08
September	27.50	88.00	48.00	592.06
October	29.50	82.00	19.00	140.27
November	20.80	97.00	20.00	52.15
December	17.90	81.00	21.00	59.332
January	19.60	81.00	27.00	105.29
February	22.40	74.00	30.00	127.12
Average ±Sd	25.52±4.27	62.50±23.23	53.83±32.49	308.53±250.33

SD ±: Standard Deviation.

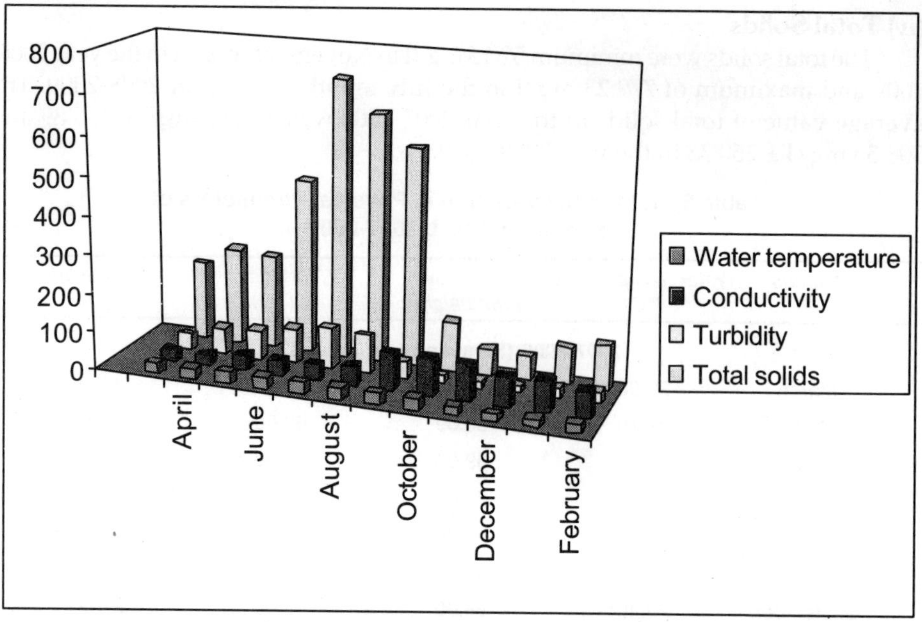

Figure 5.I.1: Graph Showing the Monthly Variation in Physical Characteristics at Sampling Site–I (2007-2008)

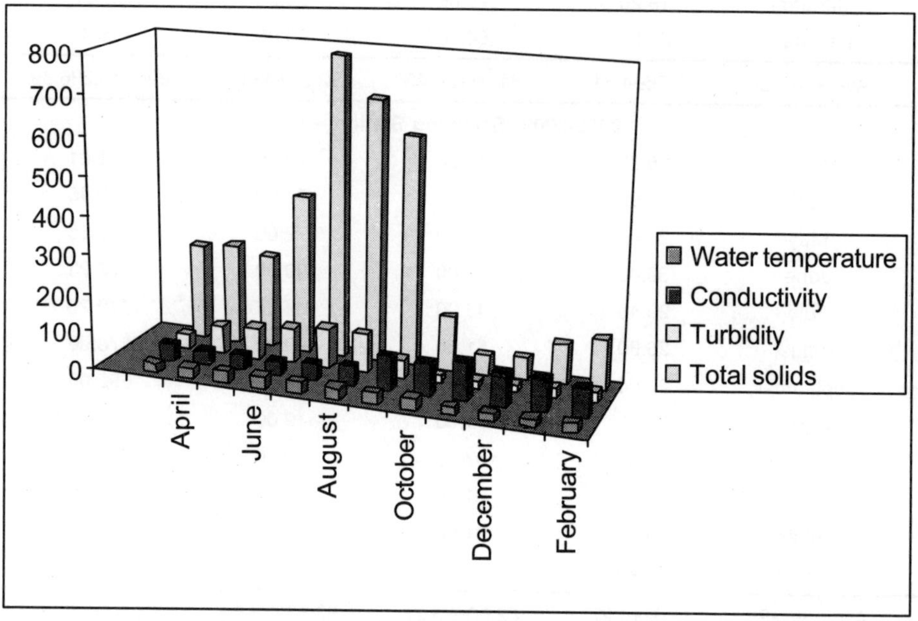

Figure 5.I.2: Graph Showing the Monthly Variation in Physical Characteristics at Sampling Site–I (2008-2009)

Sampling Site–II

The physical characteristics of sampling site–II are presented in Table 5.I.2 and Figures 5.I.3 and 5.I.4

Table 5.I.2: Monthly Variation in Physical Parameters of Sampling Site–II, 2007-2009

Months	*Water Temperature (°C)*	*Conductivity (μmhos/cm²)*	*Turbidity (JTU)*	*Total Solids (mg/l)*
		2007-2008 (Sampling Station – II)		
March	25.40	35.00	47.00	254.02
April	28.00	43.00	75.00	249.51
May	29.70	48.00	79.00	211.21
June	30.00	49.00	98.00	569.35
July	30.20	50.00	132.00	886.36
August	26.80	62.00	119.00	731.45
September	28.40	35.00	23.00	770.00
October	28.40	77.00	19.00	245.00
November	21.10	89.00	18.00	137.00
December	18.40	75.00	22.00	87.61
January	17.90	78.00	29.00	50.12
February	22.80	59.00	31.00	57.00
Average ±Sd	25.59±4347	58.33±17.98	57.67±41.41	354.05±301.07
		2008-2009 (Sampling Station – II)		
March	25.60	33.00	80.00	167.22
April	27.60	40.00	86.00	138.84
May	28.90	43.00	89.00	273.22
June	29.60	51.00	105.00	510.60
July	30.00	47.00	138.00	881.20
August	27.00	59.00	130.00	804.00
September	28.20	41.00	29.00	742.12
October	29.40	67.00	21.00	230.11
November	21.60	79.00	24.00	123.42
December	18.20	70.00	28.00	65.14
January	17.60	81.00	39.00	48.91
February	22.90	63.00	48.00	68.93
Average ±Sd	25.55±4.43	56.17±15.99	68.08±42.14	337.81±311.37

SD ±: Standard Deviation.

(*i*) Water Temperature

The water temperature at this sampling site was observed minimum of 17.60 °C in the month of January for year 2008-2009 and maximum of 30.20 °C in July in the

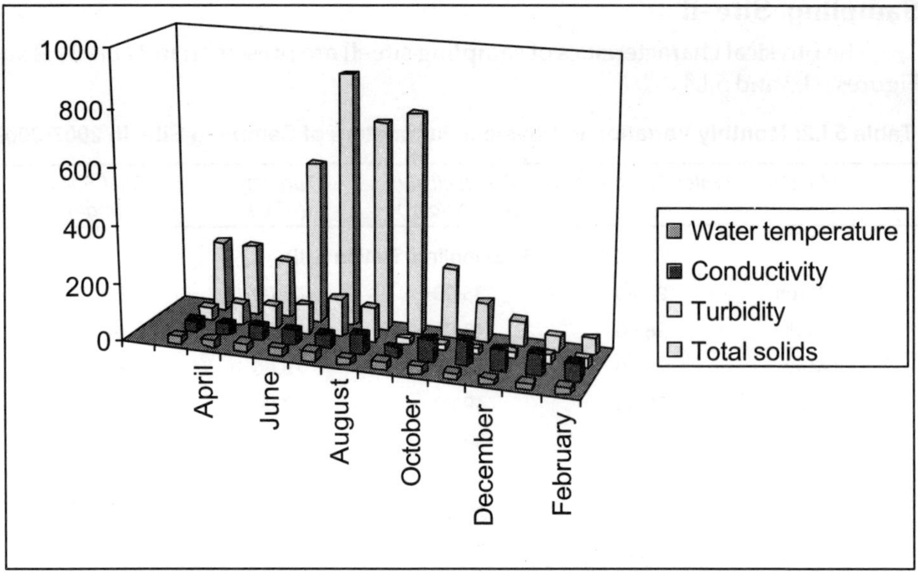

Figure 5.I.3: Graph Showing the Monthly Variation in Physical Characteristics at Sampling Site–II (2007-2008)

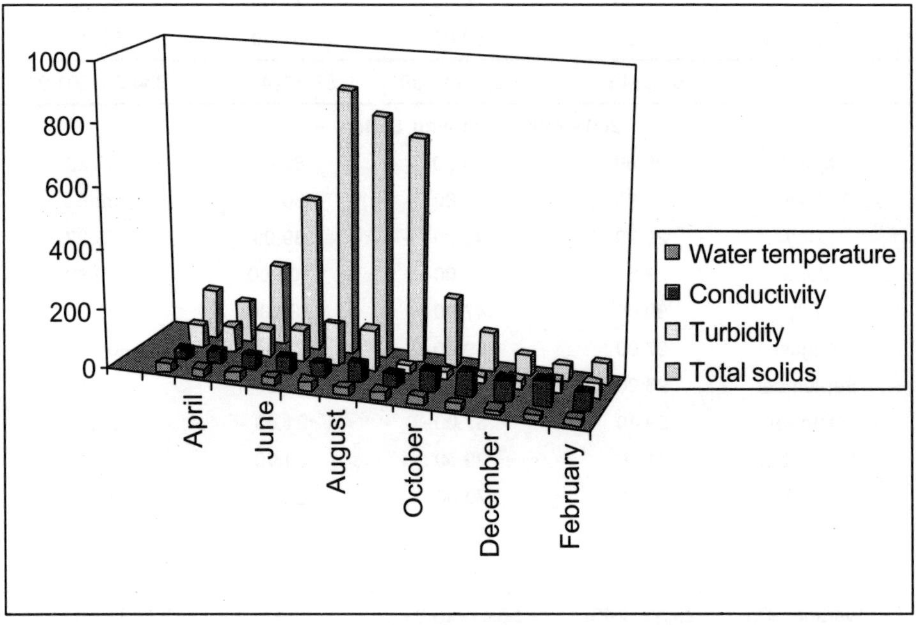

Figure 5.I.4: Graph Showing the Monthly Variation in Physical Characteristics at Sampling Site–II (2008-2009)

year 2007-2008. The average value in the year 2007-2008 was found to be 25.59 °C ± 4.37 and 25.55 °C ± 4.43 in the year 2008-2009.

(ii) Conductivity

The conductivity of pond water was observed to be minimum 33.00 μmhos/cm² in the March month in the year 2008-2009 and maximum of 89.00 μmhos/cm² in the November Month in the year 2007-2008. The average values recorded in the year2007-2008 was 58.33 μmhos/cm² ± 17.98 and 56.17 μmhos/cm² ± 15.99 in the year 2008-2009.

(iii) Turbidity

The turbidity in water was minimum 18.00 JTU in the month of November in the year 2007-2008 and maximum of 138.00 JTU in the month of July in the year 2008-2009. The average value of Turbidity in the year 2007-2008 was 57.67 JTU ± 41.41 and 68.08 JTU ± 42.14 in the year 2008-2009.

(iv) Total Solids

The total solids were minimum 48.91 mg/l in January month in the year 2008-2009 and maximum of 881.20 mg/l in the July month in the year 2008-2009. The average value of total solids in the year 2007-2008 was 34.05 mg/l ± 301.07 and 337.81 mg/l ± 311.37 in the year 2008-2009.

Sampling Site–III

The physical characteristics of sampling site–III are presented in Table 5.I.3 and Figures 5.I.5 and 5.I.6.

(i) Water Temperature

The water temperature at this sampling site was observed minimum of 17.20 °C in the month of December for year 2007-2008 and maximum of 30.00 °C in September

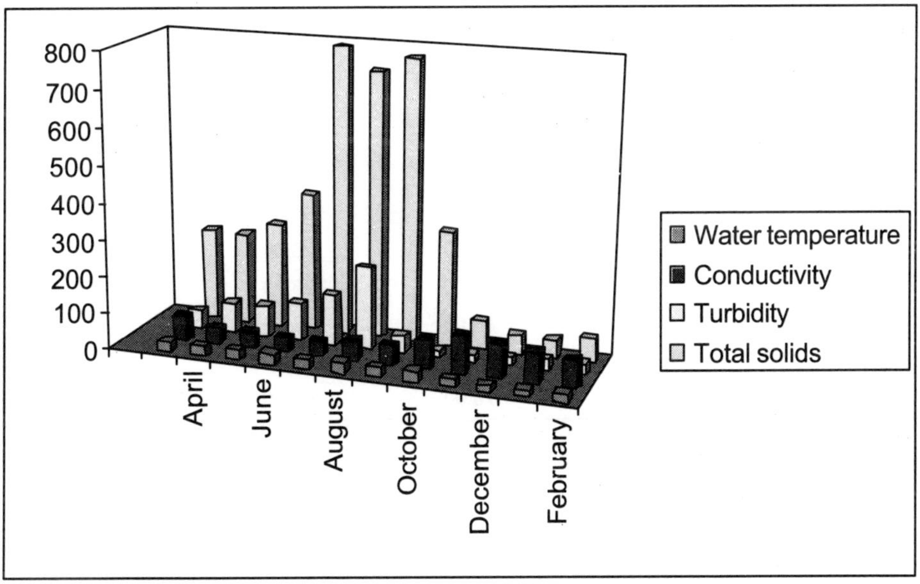

Figure 5.I.5: Graph Showing the Monthly Variation in Physical Characteristics at Sampling Site–III (2007-2008)

in the year 2008-2009. The average value in the year 2007-2008 was found to be 25.81 °C ± 4.34 and 25.67 °C ± 4.39 in the year 2008-2009.

Table 5.I.3: Monthly Variation in Physical Parameters of Sampling Site–III, 2007-2009

Months	*Water Temperature (°C)*	*Conductivity (μmhos/cm²)*	*Turbidity (JTU)*	*Total Solids (mg/l)*
		2007-2008 (Sampling Station – III)		
March	26.00	66.00	50.00	255.00
April	27.90	44.00	82.00	254.00
May	29.90	40.00	86.00	289.50
June	29.20	38.00	104.00	382.96
July	26.40	40.00	142.00	789.36
August	28.80	54.00	226.00	728.82
September	29.90	55.00	48.00	767.00
October	29.80	80.00	20.00	315.21
November	22.30	103.00	19.00	78.11
December	17.20	94.00	23.00	51.36
January	19.90	82.00	31.00	47.12
February	22.40	74.00	32.00	63.04
Average ±Sd	25.81±4.34	64.17±22.39	71.92±61.93	335.12±281.10
		2008-2009 (Sampling Station – III)		
March	25.50	70.00	79.00	273.04
April	27.70	39.00	84.00	277.41
May	29.70	36.00	94.00	297.23
June	28.20	32.00	114.00	385.09
July	29.40	39.00	148.00	808.95
August	26.90	47.00	128.00	831.00
September	30.00	50.00	54.00	701.49
October	29.90	76.00	23.00	133.71
November	21.80	91.00	28.00	64.09
December	17.80	84.00	33.00	59.01
January	19.20	87.00	42.00	49.18
February	21.90	78.00	48.00	41.68
Average ±Sd	25.67±4.39	60.75±22.26	72.92±41.56	326.82±297.42

SD ±: Standard Deviation,

(*ii*) Conductivity

The conductivity of pond water was observed to be minimum 32.00 μmhos/cm² in the June month in the year 2008-2009 and maximum of 103.00 μmhos/cm² in the

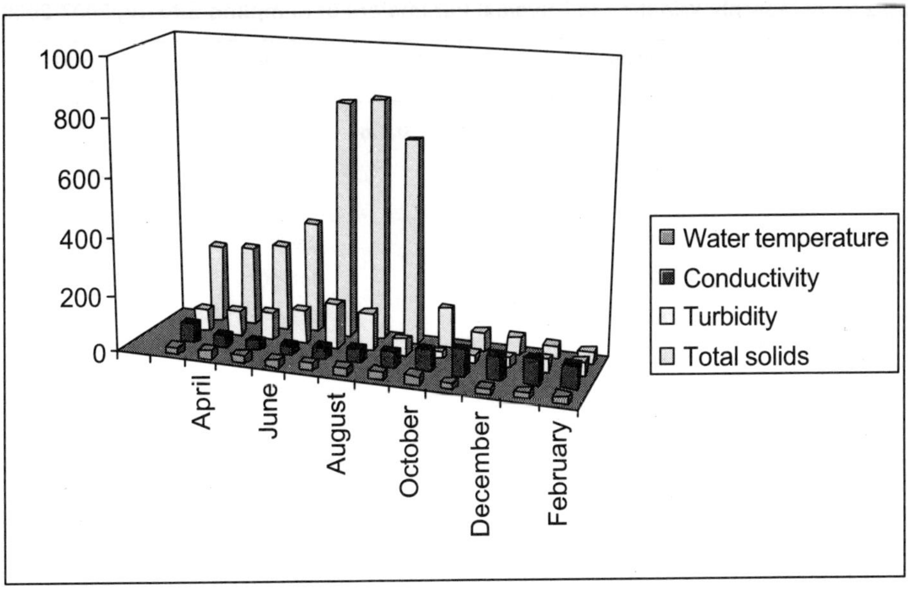

Figure 5.I.6: Graph Showing the Monthly Variation in Physical Characteristics at Sampling Site–III (2008-2009)

November Month in the year 2007-2008. The average values recorded in the year 2007-2008 was 64.17 μmhos/cm^2 ± 22.39 and 60.75 μmhos/cm^2 ± 22.26 in the year 2008-2009.

(iii) Turbidity

The turbidity in water was minimum 19.00 JTU in the month of November in the year 2007-2008 and maximum of 226.00 JTU in the month of August in the year 2007-2008. The average value of Turbidity in the year 2007-2008 was 71.92 JTU ± 61.93 and 72.93 JTU ± 41.56 in the year 2008-2009.

(iv) Total Solids

The total solids were minimum 47.12 mg/l in January month in the year 2007-2008 and maximum of 831.00 mg/l in the August month in the year 2008-2009. The average value of total solids in the year 2007-2008 was 335.12 mg/l± 281.10 and 326.82 mg/l ± 297.42 in the year 2008-2009.

Sampling Site–IV

The physical characteristics of sampling site–IV are presented in Table 5.I.4 and Figures 5.I.7 and 5.I.8.

(i) Water Temperature

The water temperature at this sampling site was observed minimum of 17.80 °C in the month of December for both the years 2007-2008 and 2008-2009 and maximum of 30.20 °C in July month in 2007-2008 and May month in the year 2008-2009. The average value in the year 2007-2008 was found to be 25.94 °C ± 4.22 and 26.04 °C ± 4.13 in the year 2008-2009.

Table 5.I.4: Monthly Variation in Physical Parameters of Sampling Site–IV, 2007-2009

Months	*Water Temperature (°C)*	*Conductivity (µmhos/cm²)*	*Turbidity (JTU)*	*Total Solids (mg/l)*
		2007-2008 (Sampling Station – IV)		
March	26.10	30.00	54.00	174.47
April	27.80	35.00	87.00	192.00
May	29.90	35.00	89.00	261.02
June	29.10	34.00	110.00	341.31
July	30.20	32.00	165.00	801.02
August	27.20	41.00	120.00	839.26
September	28.60	36.00	31.00	792.75
October	29.80	75.00	22.00	361.64
November	21.30	89.00	21.00	191.47
December	17.80	83.00	24.00	152.68
January	20.80	55.00	31.00	186.00
February	22.70	40.00	34.00	141.63
Average ±Sd	25.94±4.22	48.75±21.42	65.67±47.64	369.60±275.10
		2008-2009 (Sampling Station – IV)		
March	26.40	31.00	89.00	161.82
April	28.00	35.00	97.00	181.71
May	30.20	36.00	92.00	219.66
June	29.00	31.00	114.00	302.33
July	29.70	29.00	166.00	878.45
August	27.10	38.00	124.00	945.18
September	28.80	35.00	35.00	808.89
October	29.80	61.00	28.00	451.41
November	21.80	78.00	29.00	156.49
December	17.80	84.00	40.00	156.28
January	21.00	51.00	47.00	68.19
February	22.90	37.00	67.00	46.81
Average ±Sd	26.04±4.13	45.50±18.93	77.33±43.80	364.77±327.67

SD ±: Standard Deviation.

(ii) Conductivity

The conductivity of pond water was observed to be minimum 29.00 µmhos/cm² in the July month in the year 2008-2009 and maximum of 89.00 µmhos/cm² in the November Month in the year 2007-2008. The average values recorded in the year 2007-2008 was 48.75 µmhos/cm² ± 21.42 and 45.50 µmhos/cm² ± 18.93 in the year 2008-2009.

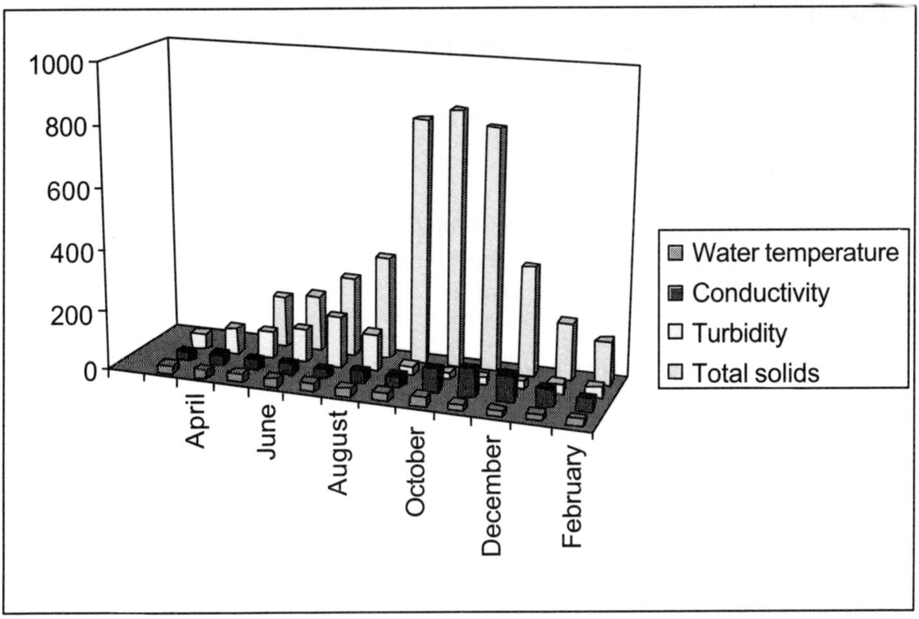

Figure 5.I.7: Graph Showing the Monthly Variation in Physical Characteristics at Sampling Site–IV (2007-2008)

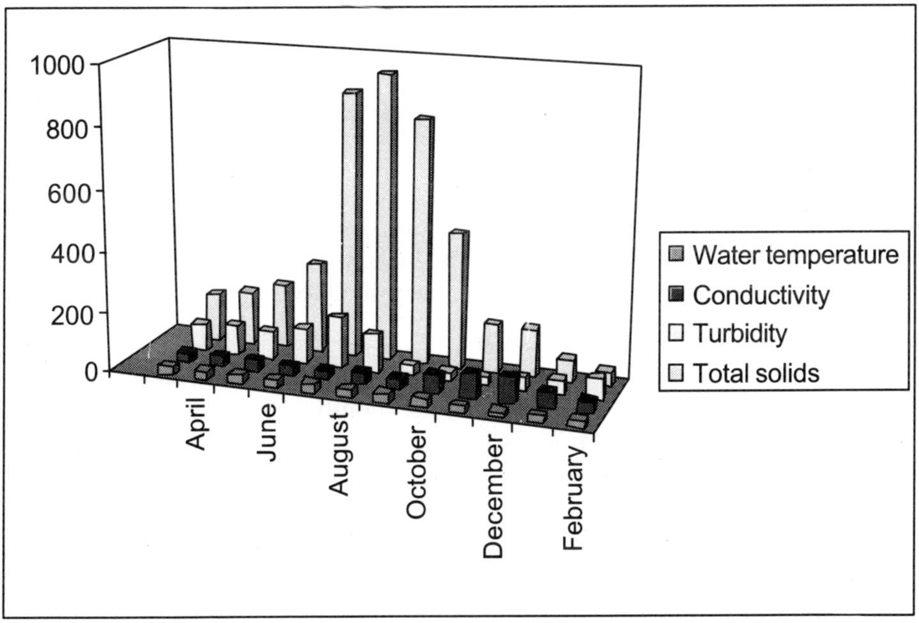

Figure 5.I.8: Graph Showing the Monthly Variation in Physical Characteristics at Sampling Site–IV (2008-2009)

(iii) Turbidity

The turbidity in water was minimum 21.00 JTU in the month of November in the year 2007-2008 and maximum of 166.00 JTU in the month of July in the year 2008-2009. The average value of Turbidity in the year 2007-2008 was 65.67 JTU ± 47.64 and 77.33 JTU ± 43.80 in the year 2008-2009.

(iv) Total Solids

The total solids were minimum 46.81 mg/l in February month in the year 2008-2009 and maximum of 945.18 mg/l in the August month in the year 2008-2009. The average value of total solids in the year 2007-2008 was 369.60 mg/l± 275.10 and 364.77 mg/l ± 327.67 in the year 2008-2009.

CHEMICAL CHARACTERISTICS

Sampling Site–I

The chemical characteristics of sampling site–I are presented in Table 5.II.1 and Figures 5.II.1 to 5.II.5.

(i) pH

pH of water was found minimum 7.94 in the August month in 2007-2008 and maximum of 8.29 in the October and May month of the year 2007-2008. The average value of pH in the year 2007-2008 was observed to be 8.20 ± 0.11 and 8.19 ± 0.10 in the year 2008-2009.

(ii) Dissolved Oxygen (DO)

Dissolved oxygen (DO) in water was found minimum 5.50 mg/l in the May month in 2007-2008 and June month of 2008-2009 and maximum of 9.30 mg/l in the December month in the year 2007-2008. The average value of DO in the year 2007-2008 was observed to be 6.92 mg/l ± 1.98 and 6.88 mg/l ± 1.27 in the year 2008-2009.

(iii) Biochemical Oxygen Demand (BOD)

BOD in water was found minimum 2.80 mg/l in the December month of both years *i.e.* 2007-2008 and 2008-2009 and maximum of 4.40 mg/l in the June month in the year 2007-2008. The average value of BOD in the year 2007-2008 was observed to be 3.53 mg/l ± 0.55 and 3.38 mg/l ± 0.47 in the year 2008-2009.

(iv) Chemical Oxygen Demand (COD)

COD in water was found minimum 7.50 mg/l in the December month in 2008-2009 and maximum of 11.30 mg/l in the June month in the year 2008-2009. The average value of COD in the year 2007-2008 was observed to be 9.90 mg/l ± 0.73 and 9.75 mg/l ± 1.25 in the year 2008-2009.

(v) Free CO_2

Free CO_2 of pond water was found to be minimum 1.00 mg/l in the months of December in the year 2007-2008 and maximum of 3.80 mg/l in the month of May in the year 2008-2009.The average value of free CO_2 in the year 2007-2008 was observed 1.96 mg/l ± 0.78 and 2.44 mg/l ± 0.94 in the year 2008-2009.

Table 5.II.1: Monthly Variation in Chemical Parameters of Sampling Site–I, 2007-2009

Months	pH	DO (mg/l)	BOD (mg/l)	COD (mg/l)	Free CO_2 (mg/l)	Total Alkalinity (mg/l)	Hardness (mg/l) Ca	Hardness (mg/l) Mg	Chlorides (mg/l)
2007-2008 (Sampling Station – I)									
March	8.26	7.40	3.60	10.30	1.43	389.00	29.70	15.64	19.50
April	8.28	6.10	3.80	10.50	2.23	370.00	30.40	16.10	24.16
May	8.29	5.50	4.30	10.70	3.60	352.00	33.50	22.3	26.98
June	8.26	5.70	4.40	11.10	2.60	334.00	34.80	23.8	29.45
July	8.04	5.90	4.00	10.60	2.30	274.00	28.40	10.80	29.60
August	7.94	5.80	3.80	9.70	2.45	310.00	27.10	9.78	19.45
September	8.10	6.20	3.50	9.10	2.23	330.00	25.90	10.09	17.80
October	8.29	6.50	2.90	9.10	2.12	442.00	27.40	11.4	17.40
November	8.27	8.40	3.00	9.00	1.01	326.00	26.50	10.9	16.00
December	8.25	9.30	2.80	9.20	1.00	352.00	26.10	12.50	15.40
January	8.22	8.20	3.00	9.50	1.30	344.00	25.70	13.29	14.90
February	8.24	8.00	3.20	9.90	1.28	367.00	26.20	14.15	17.90
Average±Sd	8.20±0.11	6.92±1.98	3.53±0.55	9.90±0.73	1.96±0.78	349.17±41.91	28.48±3.05	14.23±4.61	20.71±5.40

Contd...

Table 5.II.1–Contd...

Months	*pH*	*DO (mg/l)*	*BOD (mg/l)*	*COD (mg/l)*	*Free CO_2 (mg/l)*	*Total Alkalinity (mg/l)*	*Hardness (mg/l)*		*Chlorides (mg/l)*
							Ca	*Mg*	
2008-2009 (Sampling Station – I)									
March	8.22	7.30	3.50	10.50	3.00	402.00	29.00	15.05	19.80
April	8.25	6.00	3.70	10.80	2.50	382.00	30.90	16.18	24.54
May	8.28	5.70	4.00	11.00	3.80	361.00	33.80	22.80	27.20
June	8.24	5.50	4.30	11.30	3.50	329.00	34.90	23.60	29.40
July	8.10	5.90	3.70	10.90	2.50	297.00	28.80	11.80	29.80
August	7.94	5.60	3.20	9.60	3.50	278.00	26.80	10.20	19.20
September	8.05	6.40	3.10	9.30	1.25	303.00	25.80	9.90	18.00
October	8.27	6.40	3.00	8.80	1.15	376.00	27.90	10.80	16.70
November	8.25	8.50	2.90	7.80	1.24	331.00	27.30	9.90	15.60
December	8.23	9.20	2.80	7.50	2.56	359.00	26.80	12.20	15.00
January	8.20	8.10	3.00	9.40	1.56	361.00	26.00	12.80	14.60
February	8.19	7.90	3.40	10.10	2.70	378.00	26.70	14.20	18.60
Average±Sd	8.19±0.10	6.88±1.27	3.38±0.47	9.75±1.25	2.44±0.94	346.42±38.57	28.73±3.00	14.12±4.70	20.70±5.58

SD ±: Standard Deviation.

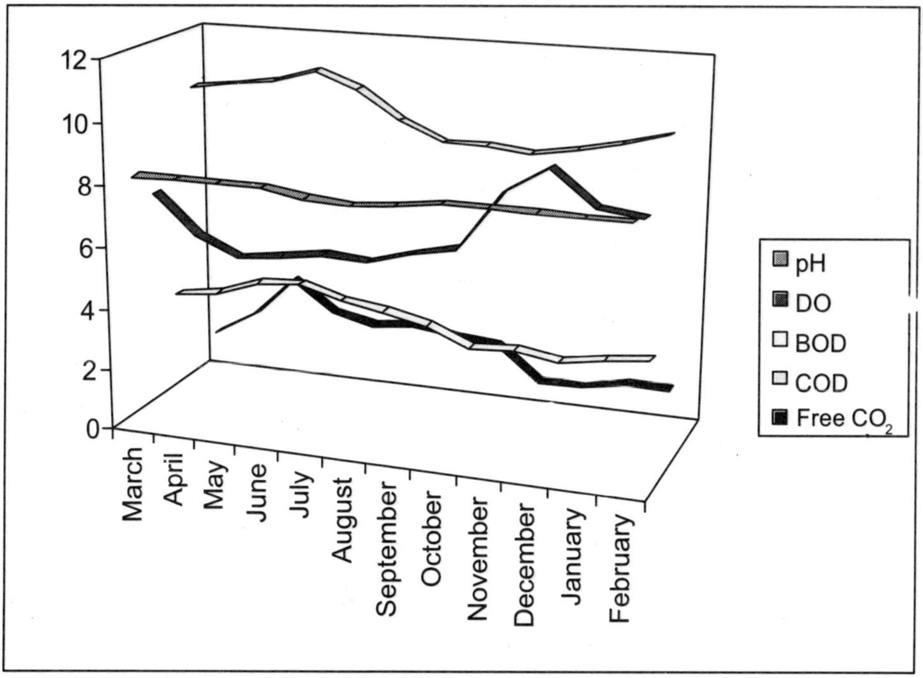

Figure 5.II.1: Graph Showing the Monthly Variation in pH, DO, BOD,COD and Free CO_2 at Sampling Site–I (2007-2008)

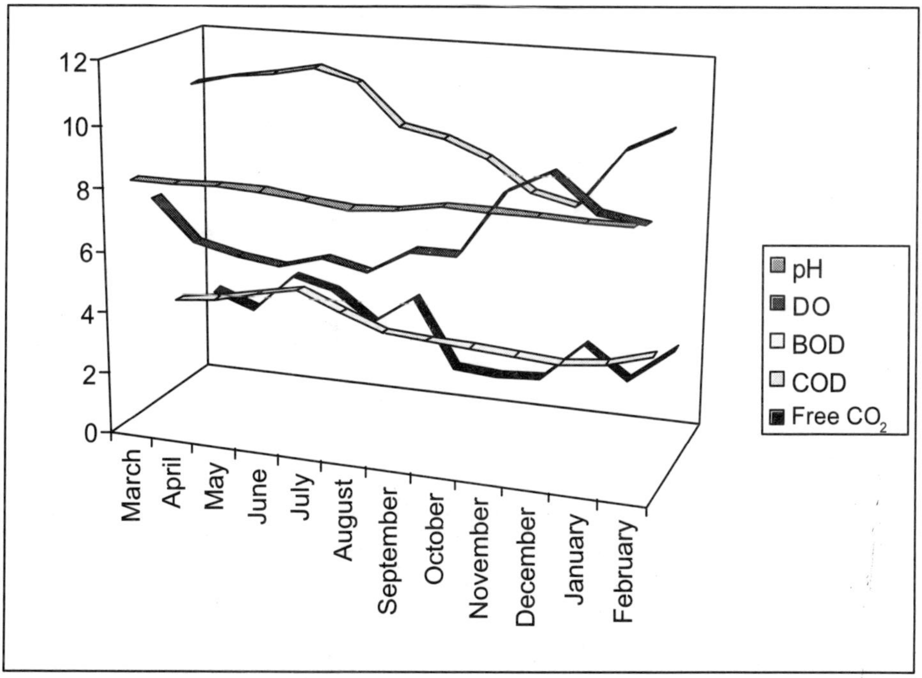

Figure 5.II.2: Graph Showing the Monthly Variation in pH, DO, BOD,COD and Free CO_2 at Sampling Site–I (2008-2009)

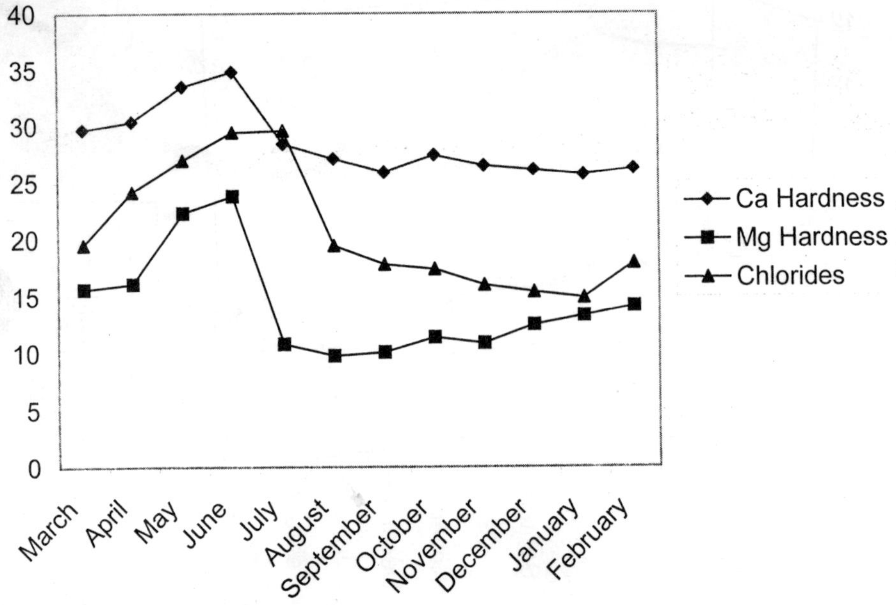

Figure 5.II.3: Graph Showing the Monthly Variation in Ca Hardness, Mg Hardness and Chlorides at Sampling Site–I (2007-2008)

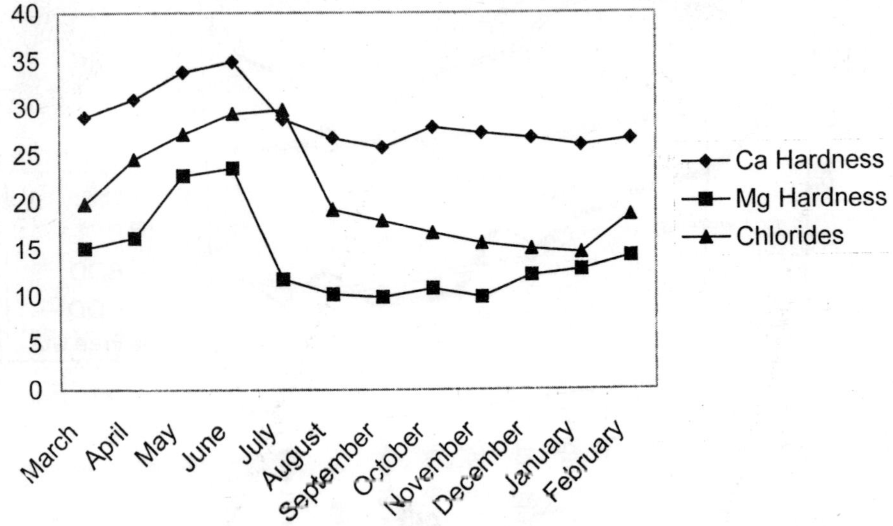

Figure 5.II.4: Graph Showing the Monthly Variation in Ca Hardness, Mg Hardness and Chlorides at Sampling Site–I (2008-2009)

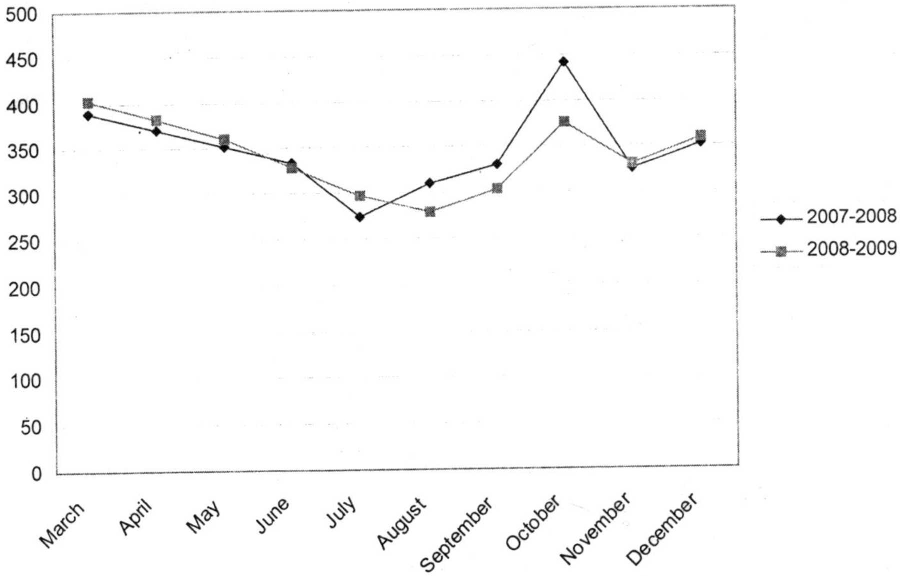

Figure 5.II.5: Graph Showing the Monthly Variation in Total Alkalinity at Sampling Site–I (2007-2009)

(vi) Total Alkalinity

The total alkalinity of pond water at this sampling station was minimum 274.00 mg/l in the month of July in the year 2007-2008 and maximum of 442.00 mg/l in the month of October in the year 2007-2008. The average value in the year 2007-2008 was 349.17 mg/l ± 41.91 and 346.42 mg/l ± 38.57 in the year 2008-2009.

(vii) Hardness

(a) Calcium Hardness

The calcium hardness in pond water at sampling site–I was minimum 25.80 mg/l in month of September in the year 2008-2009 and maximum of 34.90 mg/l in June in the year 2008-2009. The average value in the year 2007-2008 was 28.48 mg/l ± 3.05 and 28.73 mg/l ± 3.00 in the year 2008-2009.

(b) Magnesium Hardness

The magnesium hardness in pond water at sampling site–I was found minimum 9.78 mg/l in the month of August in the year 2007-2008 and maximum of 23.80 mg/l in the month of June in the year 2007-2008. The average value in the year 2007-2008 was observed 14.23 mg/l ± 4.61 and 14.12 mg/l ± 4.70 in the year 2008-2009.

(viii) Chlorides

The Chlorides in water was minimum 14.60 mg/l in the month of January in the year 2008-2009 and maximum of 29.80 mg/l in the month of July in the year 2008-2009. The average value of Chlorides in the year 2007-2008 was 20.71 mg/l ± 5.40 and 20.70 mg/l ± 5.58 in the year 2008-2009.

Sampling Site–II

The chemical characteristics of sampling site–II are presented in Table 5.II.2 and Figures 5.II.6 to 5.II.10.

(*i*) pH

pH of water was found minimum 7.85 in the August month in 2007-2008 and maximum of 8.35 in the March month in the year 2008-2009. The average value of pH in the year 2007-2008 was observed to be 8.15 ± 0.13 and 8.17± 0.17 in the year 2008-2009.

(*ii*) Dissolved Oxygen (DO)

Dissolved oxygen (DO) in water was found minimum 5.00 mg/l in the May month in 2008-2009 and maximum of 9.20 mg/l in the December month in the year 2007-2008. The average value of DO in the year 2007-2008 was observed to be 6.62 mg/l ± 1.34 and 6.65 mg/l ± 1.39 in the year 2008-2009.

(*iii*) Biochemical Oxygen Demand (BOD)

BOD in water was found minimum 2.60 mg/l in the December month of years 2007-2008 and maximum of 4.90 mg/l in the June month in the year 2008-2009. The average value of BOD in the year 2007-2008 was observed to be 3.63 mg/l ± 0.69 and 3.75 mg/l ± 0.66 in the year 2008-2009.

(*iv*) Chemical Oxygen Demand (COD)

COD in water was found minimum 8.80 mg/l in the November month in 2008-2009 and maximum of 11.90 mg/l in the June month in the year 2008-2009. The average value of COD in the year 2007-2008 was observed to be 10.31 mg/l ± 0.77 and 10.33 mg/l ± 1.01 in the year 2008-2009.

(*v*) Free CO_2

Free CO_2 of pond water was found to be minimum 1.07 mg/l in the months of December in the year 2007-2008 and maximum of 4.20 mg/l in the month of April in the year 2008-2009.The average value of free CO_2 in the year 2007-2008 was observed 2.00 mg/l ± 0.84 and 2.58 mg/l ± 0.95 in the year 2008-2009.

(*vi*) Total Alkalinity

The total alkalinity of pond water at this sampling station was minimum 206.00 mg/l in the month of June in the year 2007-2008 and maximum of 425.00 mg/l in the month of October in the year 2007-2008. The average value in the year 2007-2008 was 326.00 mg/l ± 68.40 and 321.08 mg/l ± 60.01 in the year 2008-2009.

(*vii*) Hardness

(a) Calcium Hardness

The calcium hardness in pond water at sampling site–II was minimum 26.50 mg/l in month of February in the year 2008-2009 and maximum of 38.00 mg/l in June in the year 2007-2008. The average value in the year 2007-2008 was 31.68 mg/l ± 4.12 and 29.87 mg/l ± 3.96 in the year 2008-2009.

Table 5.II.2: Monthly Variation in Chemical Parameters of Sampling Site–II, 2007-2009

Months	*pH*	*DO (mg/l)*	*BOD (mg/l)*	*COD (mg/l)*	*Free CO_2 (mg/l)*	*Total Alkalinity (mg/l)*	*Hardness (mg/l)*		*Chlorides (mg/l)*
							Ca	*Mg*	
2007-2008 (Sampling Station – II)									
March	8.18	5.70	3.80	10.60	3.26	315.00	35.10	16.80	24.60
April	8.23	5.20	3.90	11.00	3.50	289.00	36.00	17.40	27.00
May	8.29	5.50	4.20	11.30	2.35	253.00	37.90	24.50	29.30
June	8.27	5.80	4.60	11.60	2.15	206.00	38.00	26.90	31.00
July	7.95	5.40	4.50	10.80	2.54	232.00	30.10	10.50	21.20
August	7.85	5.90	4.00	10.20	2.39	382.00	28.20	11.60	29.00
September	8.15	6.50	3.90	9.90	1.39	390.00	27.90	11.90	24.00
October	8.22	6.30	2.90	9.60	1.83	425.00	28.60	12.60	18.20
November	8.19	8.70	2.70	9.20	1.13	377.00	27.80	13.40	16.50
December	8.18	9.20	2.60	9.40	1.07	355.00	27.10	14.60	17.40
January	8.16	8.10	3.10	9.70	1.15	332.00	30.50	14.80	20.00
February	8.17	7.10	3.30	10.40	1.21	356.00	33.00	16.10	21.90
Average±Sd	8.15±0.13	6.62±1.34	3.63±0.69	10.31±0.77	2.00±0.84	326.00±68.40	31.68±4.12	15.93±5.06	23.34±4.94

Contd...

Table 5.II.2–Contd...

Months	*pH*	*DO (mg/l)*	*BOD (mg/l)*	*COD (mg/l)*	*Free CO_2 (mg/l)*	*Total Alkalinity (mg/l)*	*Hardness (mg/l)*		*Chlorides (mg/l)*
							Ca	*Mg*	
2008-2009 (Sampling Station – II)									
March	8.35	5.90	3.80	11.00	2.30	321.00	28.90	16.90	20.00
April	8.29	5.60	4.00	11.20	4.20	298.00	33.50	18.80	21.20
May	8.34	5.00	4.20	11.40	3.60	262.00	37.00	25.40	26.50
June	8.27	5.30	4.90	11.90	3.10	214.00	37.80	26.50	32.00
July	7.89	5.40	4.60	10.90	2.50	241.00	28.00	10.80	33.40
August	7.79	6.00	4.20	10.50	2.70	317.00	29.00	11.20	30.00
September	8.05	6.40	3.70	9.90	1.65	370.00	28.20	11.80	26.10
October	8.24	7.10	3.10	9.40	1.32	417.00	26.90	13.00	19.30
November	8.22	8.60	3.00	8.80	1.60	381.00	27.80	11.50	17.50
December	8.20	8.90	2.80	8.90	1.54	364.00	26.90	14.30	15.80
January	8.18	8.60	3.10	9.60	3.78	321.00	27.90	14.70	17.90
February	8.21	7.00	3.60	10.40	2.70	347.00	26.50	16.40	18.10
Average±Sd	8.17±0.17	6.65±1.39	3.75±0.66	10.33±1001	2.58±0.95	321.08±60.01	29.87±3.96	15.94±5.30	23.15±6.16

SD ±: Standard Deviation.

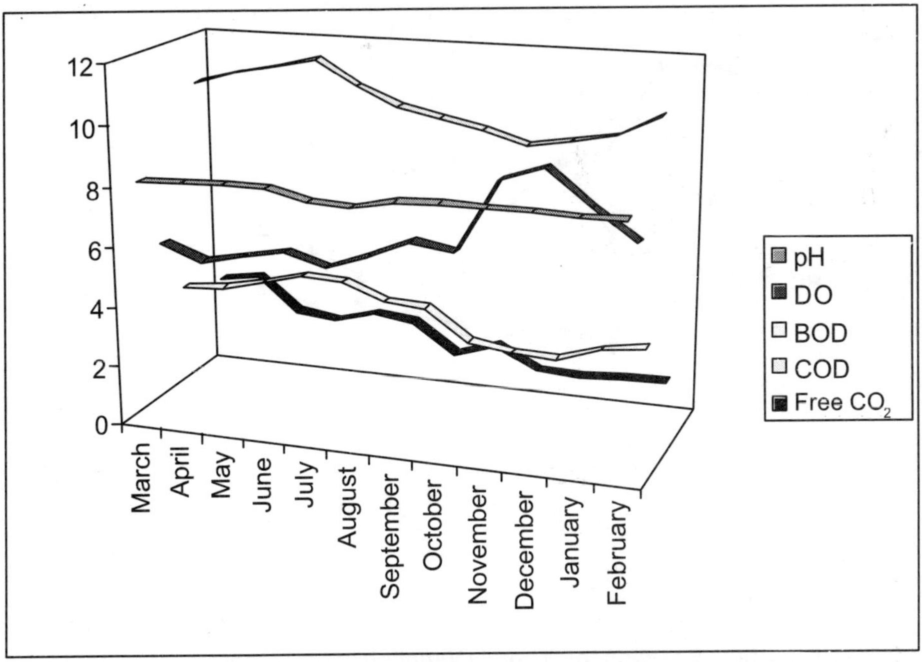

Figure 5.II.6: Graph Showing the Monthly Variation in pH, DO, BOD,COD and Free CO_2 at Sampling Site–II (2007-2008)

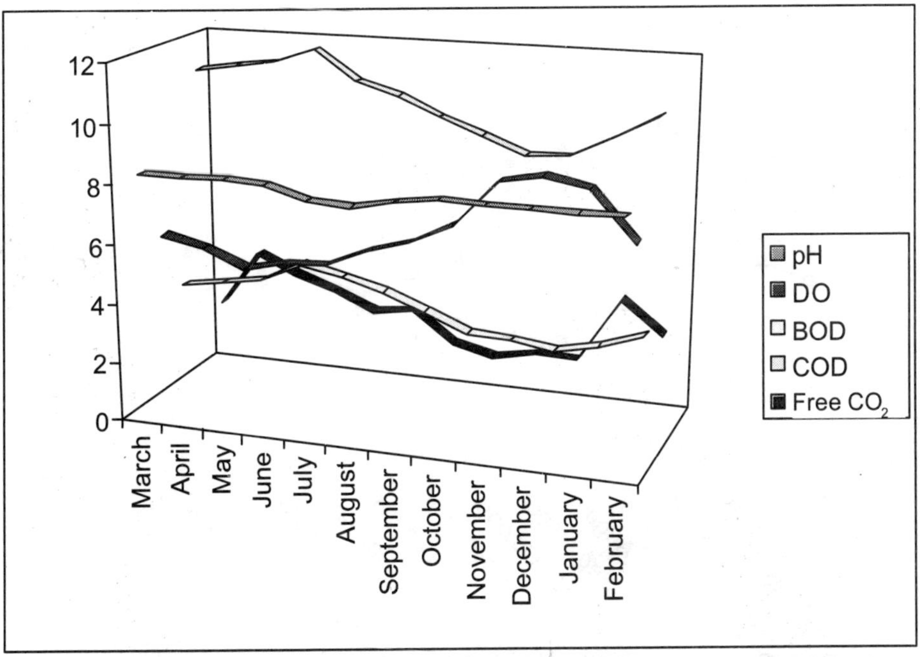

Figure 5.II.7: Graph Showing the Monthly Variation in pH, DO, BOD,COD and Free CO_2 at Sampling Site–II (2008-2009)

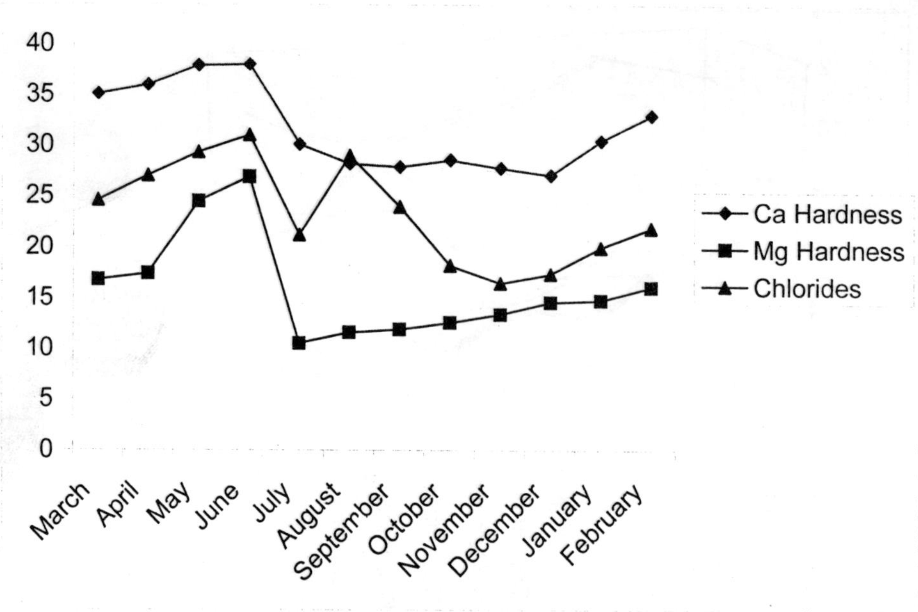

Figure 5.II.8: Graph Showing the Monthly Variation in Ca Hardness, Mg Hardness and Chlorides at Sampling Site–II (2007-2008)

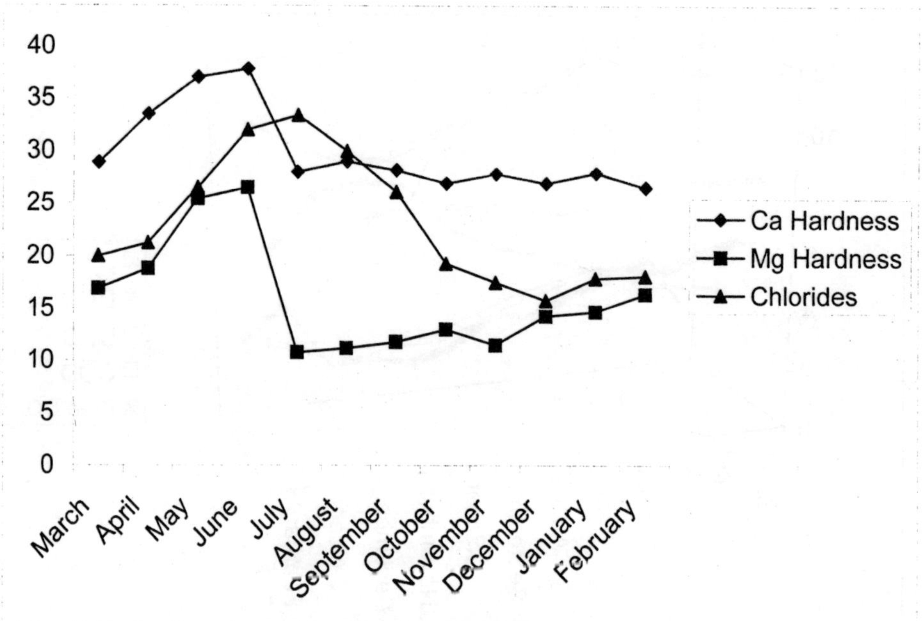

Figure 5.II.9: Graph Showing the Monthly Variation in Ca Hardness, Mg Hardness and Chlorides at Sampling Site–II (2008-2009)

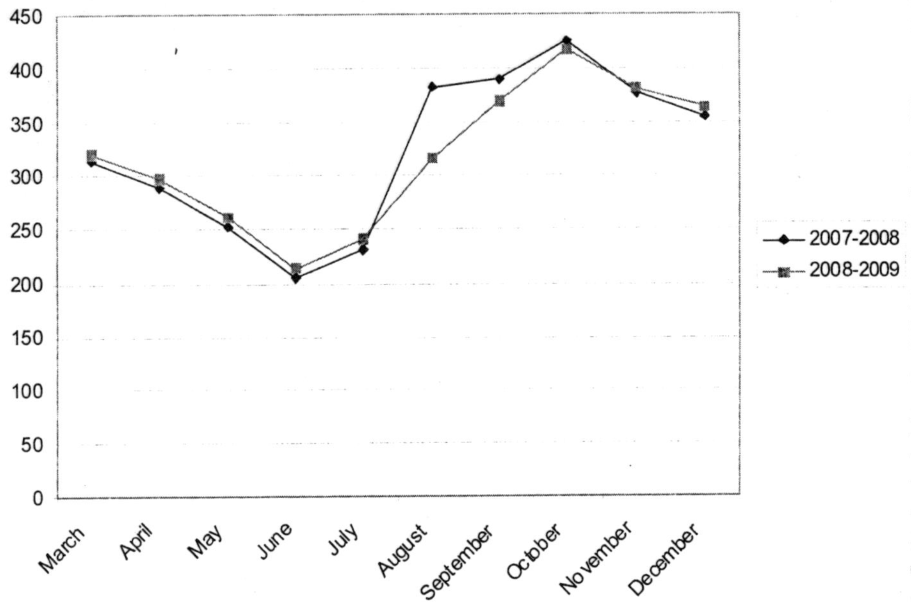

Figure 5.II.10: Graph Showing the Monthly Variation in Total Alkalinity at Sampling Site–II (2007-2009)

(b) Magnesium Hardness

The magnesium hardness in pond water at sampling site–II was found minimum 10.50 mg/l in the month of July in the year 2007-2008 and maximum of 26.90 mg/l in the month of June in the year 2007-2008. The average value in the year 2007-2008 was observed 15.93 mg/l ± 5.06 and 15.94 mg/l ± 5.30 in the year 2008-2009.

(*viii*) Chlorides

The Chlorides in water was minimum 15.80 mg/l in the month of December in the year 2008-2009 and maximum of 33.40 mg/l in the month of July in the year 2008-2009. The average value of Chlorides in the year 2007-2008 was 23.34 mg/l ± 4.94 and 23.15 mg/l ± 6.16 in the year 2008-2009.

Sampling Site–III

The chemical characteristics of sampling site-III are presented in Table 5.II.3 and Figures 5.II.11 to 5.II.15.

(*i*) pH

pH of water was found minimum 7.60 in the August month in 2008-2009 and maximum of 8.15 in the September month in the year 2008-2009. The average value of pH in the year 2007-2008 was observed to be 8.18 ± 0.16 and 8.19 ± 0.26 in the year 2008-2009.

Table 5.II.3: Monthly Variation in Chemical Parameters of Sampling Site–III, 2007-2009

Months	*pH*	*DO (mg/l)*	*BOD (mg/l)*	*COD (mg/l)*	*Free CO_2 (mg/l)*	*Total Alkalinity (mg/l)*	*Hardness (mg/l)*		*Chlorides (mg/l)*
							Ca	*Mg*	
2007-2008 (Sampling Station – III)									
March	8.25	7.0	3.7	10.9	2.30	290	30.6	17.5	27.1
April	8.27	6.1	3.9	11.2	2.10	280	34.1	18.6	28.9
May	8.30	5.2	4.2	11.5	3.30	276	36.5	23.5	29.2
June	8.28	5.4	4.9	12.1	3.10	236	39.7	27.2	33.1
July	8.00	5.5	4.8	11.6	2.71	240	31.0	12.4	27.2
August	7.75	5.3	4.6	10.9	3.07	390	29.4	12.9	19.9
September	8.20	5.9	4.2	10.3	2.37	366	29.1	13.0	20.4
October	8.25	6.5	3.2	10.2	1.85	415	30.1	12.5	17.9
November	8.22	8.6	3.1	9.7	1.13	392	29.1	13.7	16.6
December	8.23	9.0	2.9	10.1	1.09	371	28.8	14.9	16.0
January	8.21	8.2	3.0	10.4	1.53	330	28.1	15.4	20.4
February	8.22	7.7	3.4	10.7	2.07	302	27.4	16.9	23.9
Average±Sd	8.18±0.16	6.70±1.37	3.82±0.72	10.80±0.71	2.22±0.74	324.00±61.76	31.16±3.72	16.54±4.66	23.38±5.62

Contd...

Table 5.II.3–Contd...

Months	*pH*	*DO (mg/l)*	*BOD (mg/l)*	*COD (mg/l)*	*Free CO_2 (mg/l)*	*Total Alkalinity (mg/l)*	*Hardness (mg/l)*		*Chlorides (mg/l)*
							Ca	*Mg*	
2008-2009(Sampling Station – III)									
March	8.33	6.9	3.8	11.1	2.15	286	30.1	18.0	28.1
April	8.60	5.5	4.3	11.5	3.62	296	33.5	18.8	28.9
May	8.34	5.0	4.5	11.6	2.00	280	37.1	26.1	30.2
June	8.29	5.2	4.6	12.3	1.25	256	39.1	27.5	33.4
July	7.80	5.5	5.0	11.9	1.16	250	32.0	14.2	29.2
August	7.60	5.9	4.9	11.3	2.25	290	27.9	15.1	22.4
September	8.15	6.8	4.7	10.8	1.25	340	30.2	15.3	23.9
October	8.26	6.2	3.4	9.9	1.15	367	29.7	13.9	19.1
November	8.23	8.4	3.3	9.3	1.16	345	29.0	12.7	17.2
December	8.24	8.8	3.0	10.1	1.26	367	28.1	15.2	17.9
January	8.22	8.4	3.3	10.3	3.12	322	28.4	16.0	21.5
February	8.26	8.6	3.5	10.7	2.15	298	28.1	17.6	24.9
Average±Sd	8.19±0.26	6.77±1.44	4.03±0.72	10.90±0.89	1.88±0.83	308.08±39.74	31.10±3.70	17.53±4.68	24.73±5.27

SD ±: Standard Deviation.

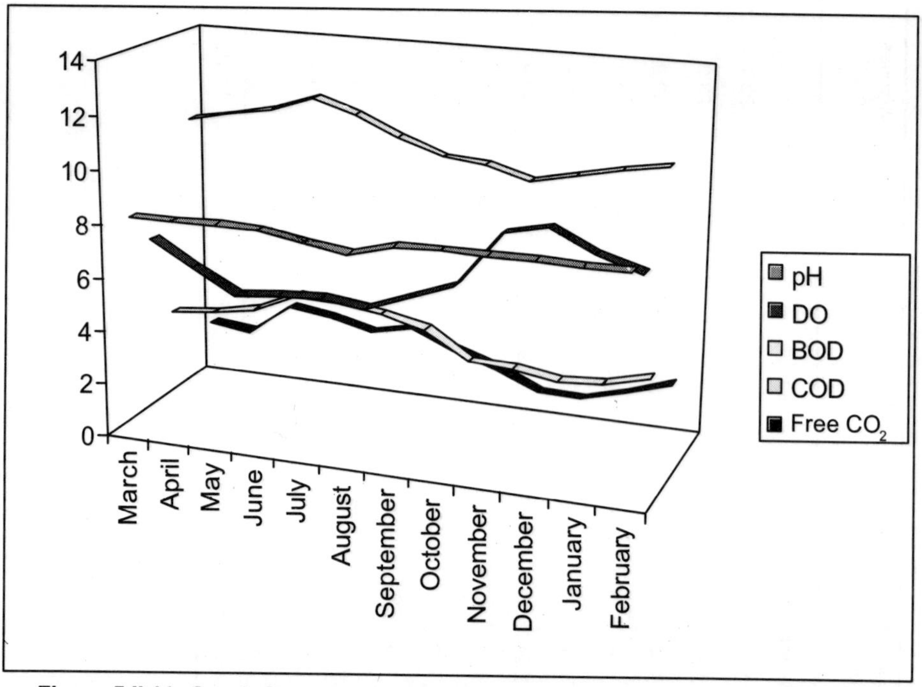

Figure 5.II.11: Graph Showing the Monthly Variation in pH, DO, BOD, COD and Free CO_2 at Sampling Site–III (2007-2008)

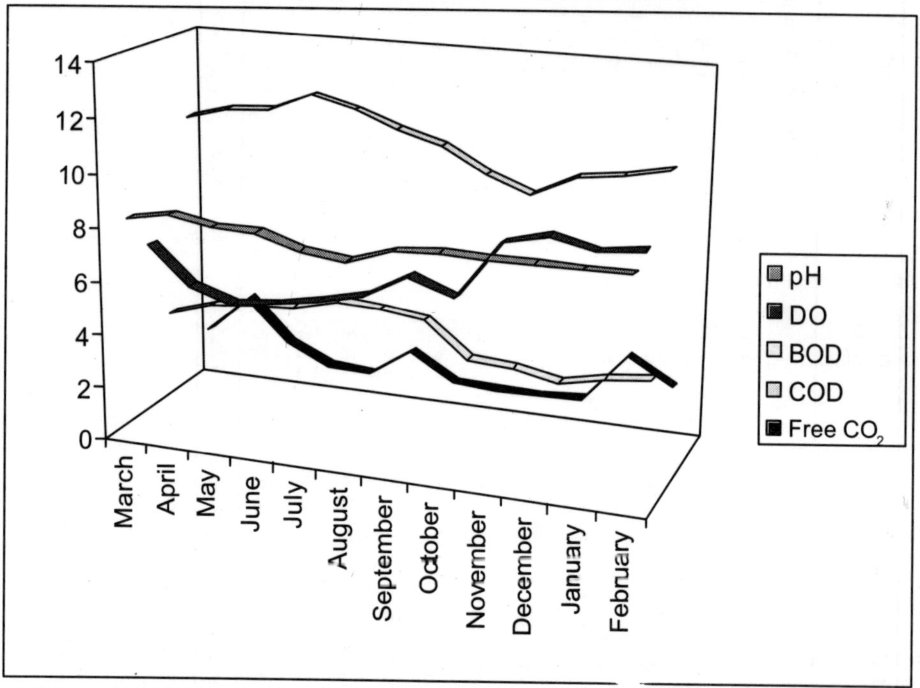

Figure 5.II.12: Graph Showing the Monthly Variation in pH, DO, BOD, COD and Free CO_2 at Sampling Site–III (2008-2009)

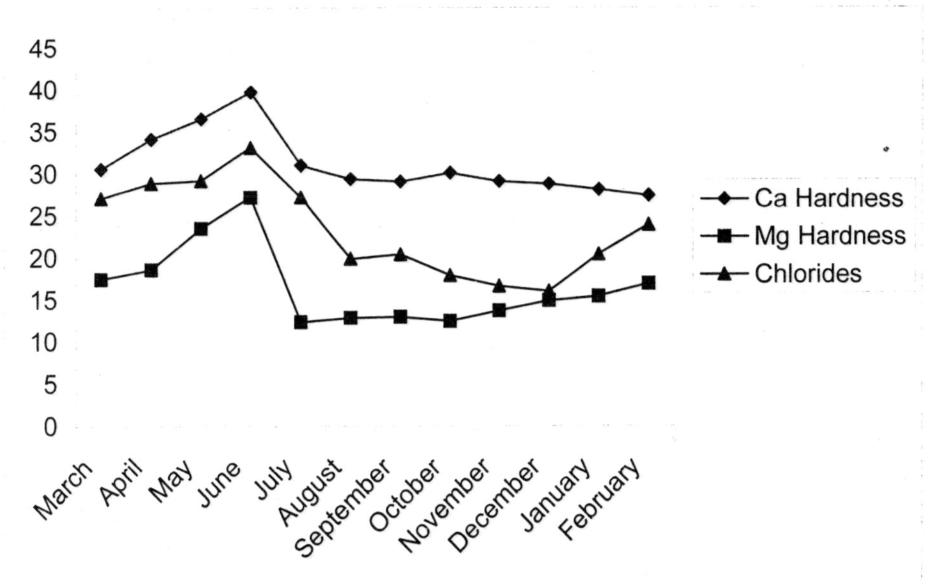

Figure 5.II.13: Graph Showing the Monthly Variation in Ca Hardness, Mg Hardness and Chlorides at Sampling Site–III (2007-2008)

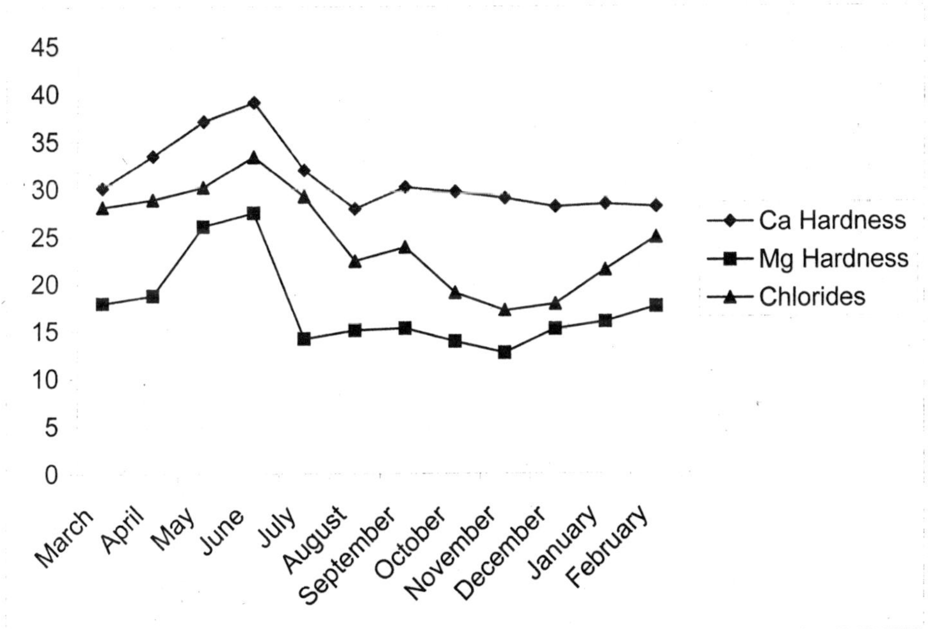

Figure 5.II.14: Graph Showing the Monthly Variation in Ca Hardness, Mg Hardness and Chlorides at Sampling Site–III (2008-2009)

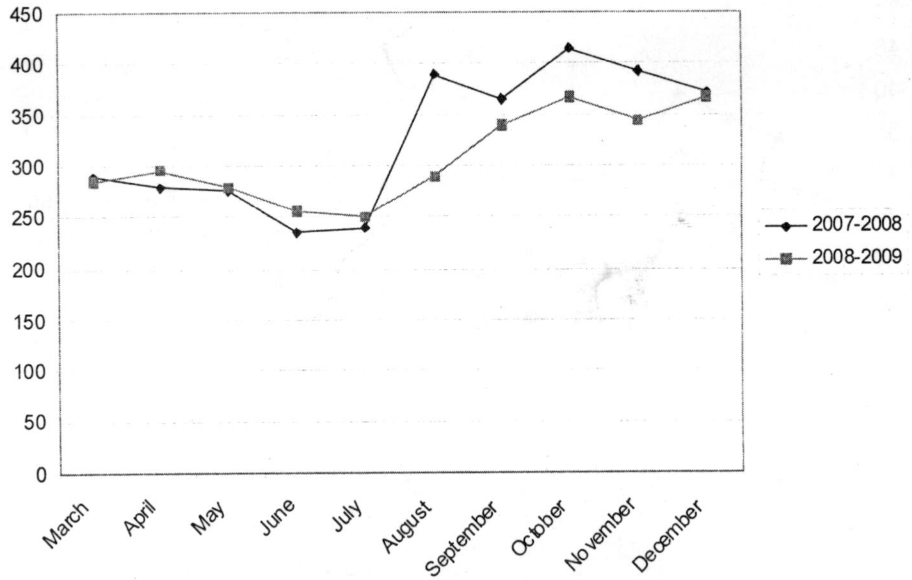

Figure 5.II.15: Graph Showing the Monthly Variation in Total Alkalinity at Sampling Site–III (2007-2009)

(*ii*) Dissolved Oxygen (DO)

Dissolved oxygen (DO) in water was found minimum 5.00 mg/l in the May month in 2008-2009 and maximum of 9.00 mg/l in the December month in the year 2007-2008. The average value of DO in the year 2007-2008 was observed to be 6.70 mg/l ± 1.37 and 6.77 mg/l ± 1.44 in the year 2008-2009.

(*iii*) Biochemical Oxygen Demand (BOD)

BOD in water was found minimum 2.90 mg/l in the December month of years 2007-2008 and maximum of 4.90 mg/l in the June month in the year 2007-2008. The average value of BOD in the year 2007-2008 was observed to be 3.82 mg/l ± 0.72 and 4.03 mg/l ± 0.72 in the year 2008-2009.

(*iv*) Chemical Oxygen Demand (COD)

COD in water was found minimum 9.30 mg/l in the November month in 2008-2009 and maximum of 12.30 mg/l in the June month in the year 2008-2009. The average value of COD in the year 2007-2008 was observed to be 10.80 mg/l ± 0.71 and 10.90 mg/l ± 0.89 in the year 2008-2009.

(*v*) Free CO_2

Free CO_2 of pond water was found to be minimum 1.09 mg/l in the months of December in the year 2007-2008 and maximum of 3.62 mg/l in the month of April in the year 2008-2009. The average value of free CO_2 in the year 2007-2008 was observed 2.22 mg/l ± 0.74 and 1.88 mg/l ± 0.83 in the year 2008-2009.

(*vi*) Total Alkalinity

The total alkalinity of pond water at this sampling station was minimum 236.00 mg/l in the month of June in the year 2007-2008 and maximum of 415.00 mg/l in the month of October in the year 2007-2008. The average value in the year 2007-2008 was 324.00 mg/l ± 61.76 and 308.08 mg/l ± 39.74 in the year 2008-2009.

(*vii*) Hardness

(a) Calcium Hardness

The calcium hardness in pond water at sampling site–III was minimum 27.40 mg/l in month of February in the year 2007-2008 and maximum of 39.70 mg/l in June in the year 2007-2008. The average value in the year 2007-2008 was 31.16 mg/l ± 3.72 and 31.10 mg/l ± 3.70 in the year 2008-2009.

(b) Magnesium Hardness

The magnesium hardness in pond water at sampling site–III was found minimum 12.40 mg/l in the month of July in the year 2007-2008 and maximum of 27.50 mg/l in the month of June in the year 2008-2009. The average value in the year 2007-2008 was observed 16.54 mg/l ± 4.66 and 17.53 mg/l ± 4.68 in the year 2008-2009.

(*viii*) Chlorides

The Chlorides in water was minimum 16.00 mg/l in the month of December in the year 2007-2008 and maximum of 33.40 mg/l in the month of June in the year 2008-2009. The average value of Chlorides in the year 2007-2008 was 23.38 mg/l ± 5.62 and 24.73 mg/l ± 5.27 in the year 2008-2009.

Sampling Site–IV

The chemical characteristics of sampling site-IV are presented in Table 5.II.4 and Figures 5.II.16 to 5.II.20.

(*i*) pH

pH of water was found minimum 7.51 in the July month in 2007-2008 and maximum of 8.38 in the May month in the year 2008-2009. The average value of pH in the year 2007-2008 was observed to be 8.16 ± 0.27 and 8.16 ± 0.29 in the year 2008-2009.

(*ii*) Dissolved oxygen (DO)

Dissolved oxygen (DO) in water was found minimum 4.20 mg/l in the May month in 2008-2009 and maximum of 8.70 mg/l in the December month in the year 2008-2009. The average value of DO in the year 2007-2008 was observed to be 6.13 mg/l ± 1.43 and 6.25 mg/l ± 1.49 in the year 2008-2009.

(*iii*) Biochemical Oxygen Demand (BOD)

BOD in water was found minimum 3.00 mg/l in the December month of year 2007-2008 and maximum of 4.90 mg/l in the June month in the year 2007-2008. The average value of BOD in the year 2007-2008 was observed to be 3.86 mg/l ± 0.65 and 3.93 mg/l ± 0.64 in the year 2008-2009.

Table 5.II.4: Monthly Variation in Chemical Parameters of Sampling Site–IV, 2007-2009

Months	*pH*	*DO (mg/l)*	*BOD (mg/l)*	*COD (mg/l)*	*Free CO_2 (mg/l)*	*Total Alkalinity (mg/l)*	*Hardness (mg/l)*		*Chlorides (mg/l)*
							Ca	*Mg*	
2007-2008 (Sampling Station – IV)									
March	8.27	6.70	3.90	10.80	2.02	362.00	30.20	17.20	26.60
April	8.29	4.90	4.10	11.20	3.23	380.00	34.30	18.30	27.00
May	8.35	4.30	4.40	11.40	3.73	342.00	36.20	25.30	28.00
June	8.30	4.60	4.90	11.90	2.56	270.00	36.80	26.20	34.20
July	7.51	5.20	4.60	11.40	2.30	264.00	28.30	11.20	22.60
August	7.64	5.10	4.40	10.70	2.45	280.00	25.60	10.90	20.20
September	8.24	5.90	4.00	10.00	2.23	320.00	28.40	11.40	19.70
October	8.26	5.30	3.30	10.10	3.23	418.00	30.20	12.20	21.20
November	8.24	7.70	3.10	9.80	1.85	294.00	28.60	14.20	18.10
December	8.27	8.20	3.00	9.90	1.55	284.00	29.30	14.80	20.20
January	8.24	8.00	3.20	10.30	1.61	282.00	26.20	15.30	21.90
February	8.25	7.60	3.40	10.70	2.01	325.00	27.40	15.90	23.40
Average±Sd	8.16±0.27	6.13±1.43	3.86±0.65	10.68±0.68	2.40±0.69	318.12±48.71	30.13±3.71	16.01±5.11	23.59±4.58

Contd...

Table 5.II.4–Contd...

Months	*pH*	*DO (mg/l)*	*BOD (mg/l)*	*COD (mg/l)*	*Free CO_2 (mg/l)*	*Total Alkaiinity (mg/l)*	*Hardness (mg/l)*		*Chlorides (mg/l)*
							Ca	*Mg*	
2008-2009 (Sampling Station – IV)									
March	8.36	6.40	3.80	11.00	3.80	342.00	30.20	17.20	26.40
April	8.33	5.10	4.20	11.20	4.20	362.00	34.00	18.40	27.90
May	8.38	4.20	4.40	11.60	3.50	333.00	36.50	25.60	29.10
June	8.32	4.50	5.10	11.40	2.50	292.00	37.20	26.60	34.40
July	7.55	5.20	4.70	12.20	1.50	269.00	28.60	12.40	23.40
August	7.58	5.30	4.40	11.20	2.50	277.00	26.00	11.20	21.70
September	8.22	6.10	3.90	10.70	1.35	298.00	27.30	11.10	20.20
October	8.29	6.20	3.60	9.80	1.25	379.00	30.60	13.20	20.30
November	8.24	8.20	3.30	9.00	1.46	398.00	29.20	10.20	18.30
December	8.26	8.70	3.10	9.40	3.50	315.00	28.20	15.30	19.80
January	8.20	7.90	3.30	10.00	3.20	291.00	27.40	14.60	22.20
February	8.23	7.30	3.30	10.50	3.60	327.00	26.30	16.90	23.60
Average±Sd	8.16±0.29	6.25±1.49	3.93±0.64	10.67±0.96	2.70±1.10	323.58±40.97	30.13±3.18	16.06±5.36	23.95±4.69

SD ±: Standard Deviation.

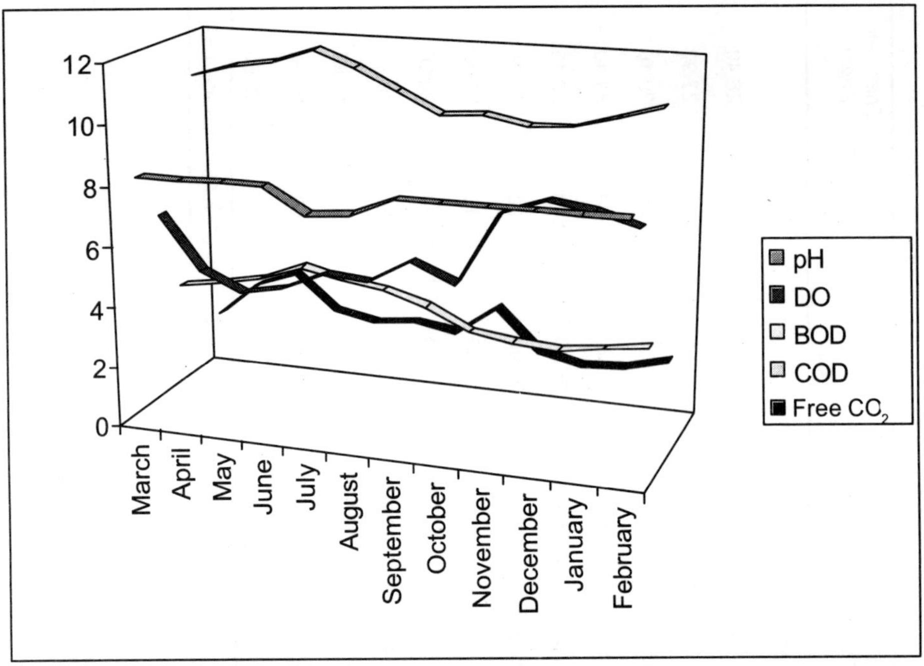

Figure 5.II.16: Graph Showing the Monthly Variation in pH, DO, BOD, COD and Free CO_2 at Sampling Site–IV (2007-2008)

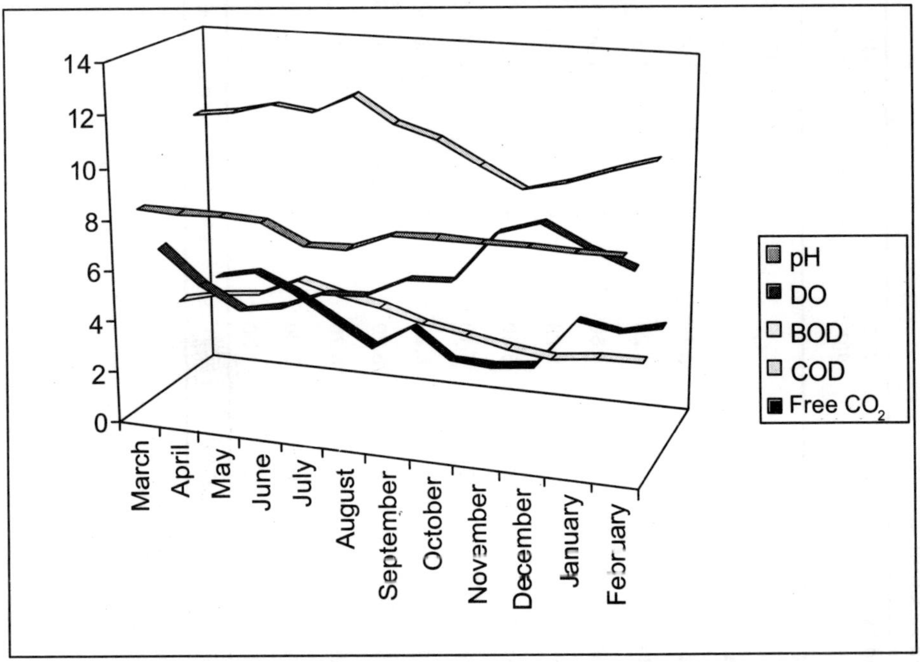

Figure 5.II.17: Graph Showing the Monthly Variation in pH, DO, BOD, COD and Free CO_2 at Sampling Site–IV (2008-2009)

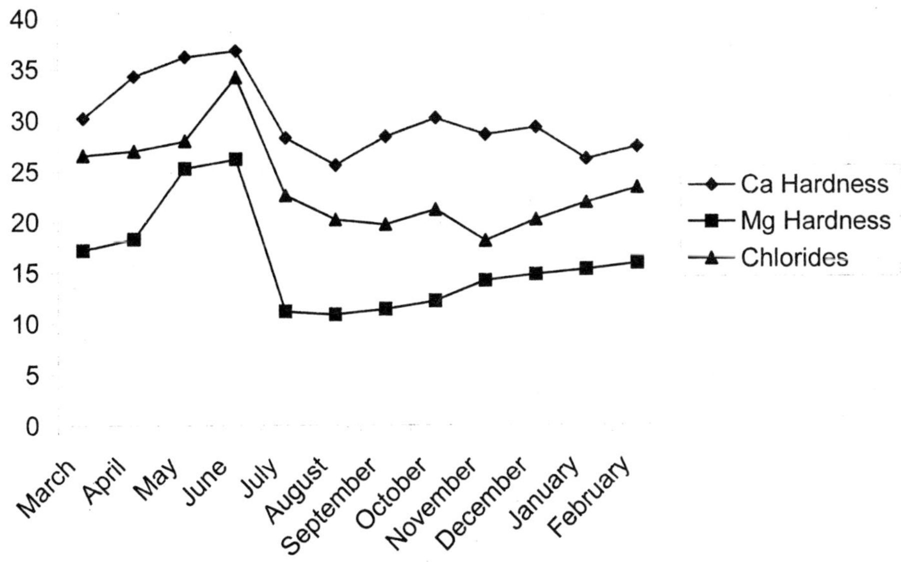

Figure 5.II.18: Graph Showing the Monthly Variation in Ca Hardness, Mg Hardness and Chlorides at Sampling Site–IV (2007-2008)

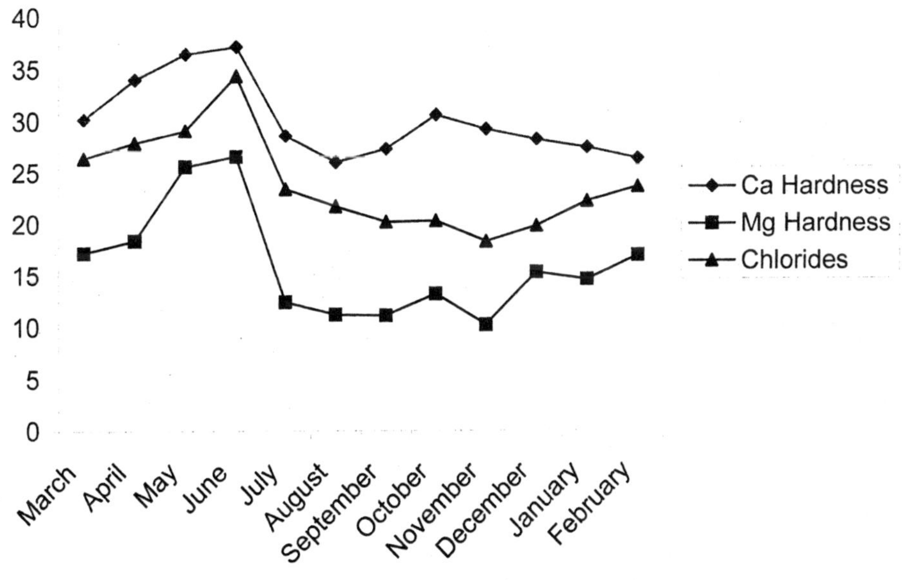

Figure 5.II.19: Graph Showing the Monthly Variation in Ca Hardness, Mg Hardness and Chlorides at Sampling Site–IV (2008-2009)

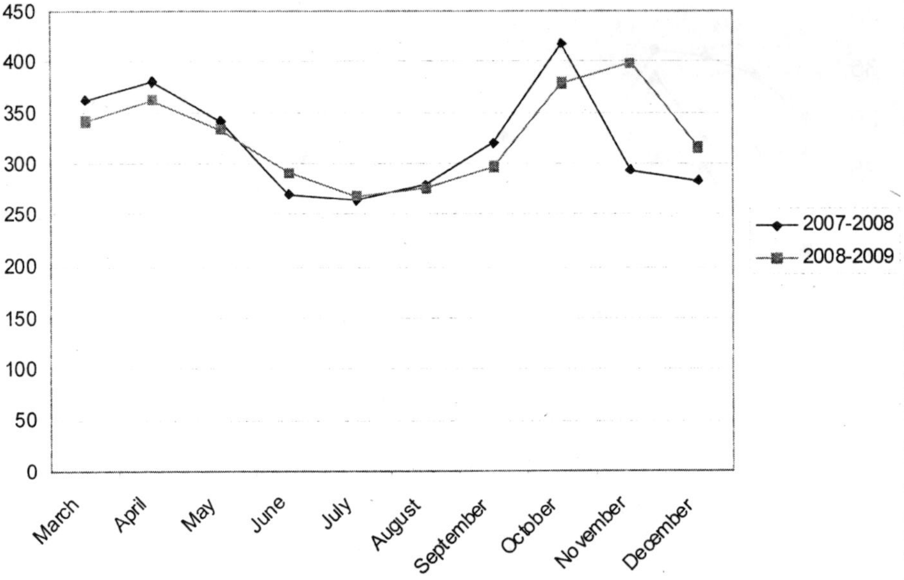

Figure 5.II.20: Graph Showing the Monthly Variation in Total Alkalinity at Sampling Site–IV (2007-2009)

(*iv*) Chemical Oxygen Demand (COD)

COD in water was found minimum 9.00 mg/l in the November month in 2008-2009 and maximum of 11.90 mg/l in the June month in the year 2007-2008. The average value of COD in the year 2007-2008 was observed to be 10.68 mg/l ± 0.68 and 10.67 mg/l ± 0.96 in the year 2008-2009.

(*v*) Free CO_2

Free CO_2 of pond water was found to be minimum 1.25 mg/l in the month of October in the year 2008-2009 and maximum of 4.20 mg/l in the month of April in the year 2008-2009.The average value of free CO_2 in the year 2007-2008 was observed 2.40 mg/l ± 0.69 and 2.70 mg/l ± 1.10 in the year 2008-2009.

(*vi*) Total Alkalinity

The total alkalinity of pond water at this sampling station was minimum 264.00 mg/l in the month of July in the year 2007-2008 and maximum of 418.00 mg/l in the month of October in the year 2007-2008. The average value in the year 2007-2008 was 318.12 mg/l ± 48.71 and 323.58 mg/l ± 40.97 in the year 2008-2009.

(*vii*) Hardness

(a) Calcium Hardness

The calcium hardness in pond water at sampling site–IV was minimum 25.60 mg/l in month of August in the year 2007-2008 and maximum of 37.20 mg/l in June in the year 2008-2009. The average value in the year 2007-2008 was 30.13 mg/l ± 3.71 and 30.13 mg/l ± 3.81 in the year 2008-2009.

(b) Magnesium Hardness

The magnesium hardness in pond water at sampling site–IV was found minimum 10.20 mg/l in the month of November in the year 2008-2009 and maximum of 26.60 mg/l in the month of June in the year 2008-2009. The average value in the year 2007-2008 was observed 16.01 mg/l ± 5.11 and 16.06 mg/l ± 5.36 in the year 2008-2009.

(*viii*) Chlorides

The Chlorides in water was minimum 18.10 mg/l in the month of November in the year 2007-2008 and maximum of 34.40 mg/l in the month of June in the year 2008-2009. The average value of Chlorides in the year 2007-2008 was 23.59 mg/l ± 4.58 and 23.95 mg/l ± 4.69 in the year 2008-2009.

BIOLOGICAL PARAMETERS

A study on the biological parameters of the Paniyala pond includes the primary producers like phytoplankters, primary consumers like zooplankters, aquatic plants and decomposers like bacteria and fungi. Site wise details of the observations of biological parameters are mentioned in tables (Table 5.III.A.1 to Table 5.III.A.12) and figures (Figure 5.III.A.1 to Figure 5.III.A.24)

A. Plankters

Sampling Site–I

(*a*) Plankters

Quantitative analysis of plankters at Sampling site–I revealed that the total plankters number per liter of water was observed maximum 3816.00 unit/l in the month of January in the year 2008-2009 and minimum was 422.00 unit/l in July month of the 2008-2009. The average value of plankters in 2007-2008 is 1836.00 unit/l ±421.93 and in 2008-2009 was 1572.33 unit/l ±224.45 (Tables 5.III.A.1 to 5.III.A.3 and Figures 5.III.A.1 to 5.III.A.6)

1. Phytoplankters

Phytoplankters was noted lowest of 404.00 unit/l in the month of July of 2008-2009 and highest of 3297.00 unit/l in January month of 2008-2009 with the average cells number 1617.33 unit/l ±280.06 in 2007-2008 and 1358.50 unit/l ±185.88 in 2008-2009.

In phytoplankters Diatoms were dominated and class blue green algae was found least in both the years of study. At this sampling site diatoms was observed maximum in the month of January of 2007-2008 (2846.00 unit/l) and minimum in July month (382.00 unit/l) of 2008-2009 with average count of cells in 2007-2008 was 1151.42 unit/l ±229.27 and 1086.88 unit/l ±130.22 in 2008-2009. Green algae were found to be highest in the month of December 2007-2008 (433.00 unit/l) and lowest in 2008-2009 in the month of July (14.00 unit/l). The average count of Green algae in 2007-2008 was 170.75 unit/l ± 33.48 and 142.58 unit/l ±44.72 in 2008-2009. Blue green algae showed optimum peak during the month of January in 2008-2009 (162.00

Table 5.III.A.1: Quantitative Analysis of the Plankters of the Paniyala Pond During 2007-2009 at Sampling Site–I

Months	*Number of Plankters per litre of Water*				*Total Plankters (Unit/l)*	
	Phytoplankters (Unit/l)		*Zooplankters (Unit/l)*			
	2007-2008	*2008-2009*	*2007-2008*	*2008-2009*	*2007-2008*	*2008-2009*
March	1322.00	1247.00	450.00	437.00	1772.00	1684.00
April	1237.00	1192.00	423.00	214.00	1660.00	1406.00
May	939.00	1138.00	274.00	220.00	1213.00	1358.00
June	681.00	654.00	96.00	46.00	777.00	700.00
July	601.00	404.00	39.00	18.00	641.00	422.00
August	821.00	674.00	48.00	24.00	869.00	698.00
September	1268.00	903.00	102.00	73.00	1370.00	976.00
October	2033.00	1067.00	146.00	120.00	2179.00	1187.00
November	2405.00	1361.00	181.00	195.00	2586.00	1556.00
December	2737.00	1894.00	212.00	262.00	2949.00	2156.00
January	3287.00	3297.00	347.00	519.00	3634.00	3816.00
February	2076.00	2470.00	306.00	438.00	2382.00	2908.00
Average±SD	1617.33±280.06	1358.50±185.88	218.67±51.87	213.83±38.57	1836.00±421.93	1572.33±224.45

SD ±: Standard Deviation.

Table 5.III.A.2: Number of Different Group Among the Phytoplankters of the Paniyala Pond During 2007-2009 at Sampling Site–I

Months	*Total Phytoplankters/ Liter of Water (Unit/l)*		*Total Diatoms/ Liter of Water (Unit/l)*		*Total Green Algae/ Liter of Water (Unit/l)*		*Total Blue Green Algae/ Liter of Water (Unit/l)*	
	2007-2008	*2008-2009*	*2007-2008*	*2008-2009*	*2007-2008*	*2008-2009*	*2007-2008*	*2008-2009*
March	1322.00	1247.00	1098.00	1048.00	184.00	162.00	40.00	36.00
April	1237.00	1192.00	975.00	1015.00	186.00	135.00	75.00	42.00
May	939.00	1138.00	836.00	1010.00	72.00	106.00	30.00	22.00
June	681.00	654.00	610.00	562.00	66.00	85.00	5.00	6.00
July	601.00	404.00	546.00	382.00	52.00	14.00	2.00	8.00
August	821.00	674.00	970.00	654.00	45.00	15.00	5.00	5.00
September	1268.00	903.00	1100.00	782.00	156.00	95.00	12.00	25.00
October	2033.00	1067.00	1820.00	897.00	182.00	125.00	30.00	45.00
November	2405.00	1362.00	1957.00	1182.00	406.00	164.00	42.00	15.00
December	2737.00	1894.00	2129.00	1621.00	433.00	216.00	75.00	56.00
January	3287.00	3297.00	2846.00	2783.00	364.00	352.00	76.00	162.00
February	2076.00	2470.00	1752.00	2106.00	242.00	242.00	82.00	121.00
Average± SD	1617.33± 280.06	1358.50± 185.88	1151.42± 229.27	1086.88± 130.22	170.75± 33.48	142.58± 44.72	36.17± 17.31	45.25± 10.94

SD ±: Standard Deviation.

Table 5.III.A.3: Number of Different Genera Among Zooplankters in the Paniyala Pond During 2007-2009 at Sampling Site–I

Months	*Protozoa (Unit/l)*		*Rotifera (Unit/l)*		*Cladocera (Unit/l)*		*Copepod (Unit/l)*		*Ostracods (Unit/l)*	
	2007-2008	*2008-2009*	*2007-2008*	*2008-2009*	*2007-2008*	*2008-2009*	*2007-2008*	*2008-2009*	*2007-2008*	*2008-2009*
March	124.00	112.00	86.00	135.00	32.00	112.00	102.00	0.00	106.00	78.00
April	108.00	46.00	106.00	120.00	15.00	32.00	98.00	0.00	96.00	16.00
May	82.00	55.00	85.00	95.00	10.00	35.00	62.00	0.00	35.00	35.00
June	52.00	1.00	15.00	15.00	2.00	15.00	12.00	0.00	15.00	15.00
July	18.00	5.00	6.00	5.00	4.00	3.00	8.00	0.00	3.00	5.00
August	25.00	6.00	8.00	8.00	1.00	5.00	12.00	0.00	2.00	5.00
September	35.00	2.00	32.00	30.00	5.00	21.00	15.00	0.00	15.00	20.00
October	20.00	15.00	62.00	45.00	15.00	25.00	24.00	0.00	25.00	35.00
November	38.00	46.00	72.00	52.00	3.00	52.00	36.00	0.00	32.00	45.00
December	40.00	95.00	56.00	86.00	8.00	16.00	62.00	0.00	46.00	65.00
January	105.00	176.00	78.00	186.00	15.00	52.00	85.00	0.00	64.00	105.00
February	85.00	106.00	64.00	156.00	26.00	78.00	106.00	0.00	25.00	98.00
Average±SD	61.00±13.53	55.41±9.86	55.65±12.07	77.75±11.43	11.33±3.22	37.17±6.19	51.83±13.44	0.00±0.00	38.67±9.61	43.50±11.10

SD ±: Standard Deviation.

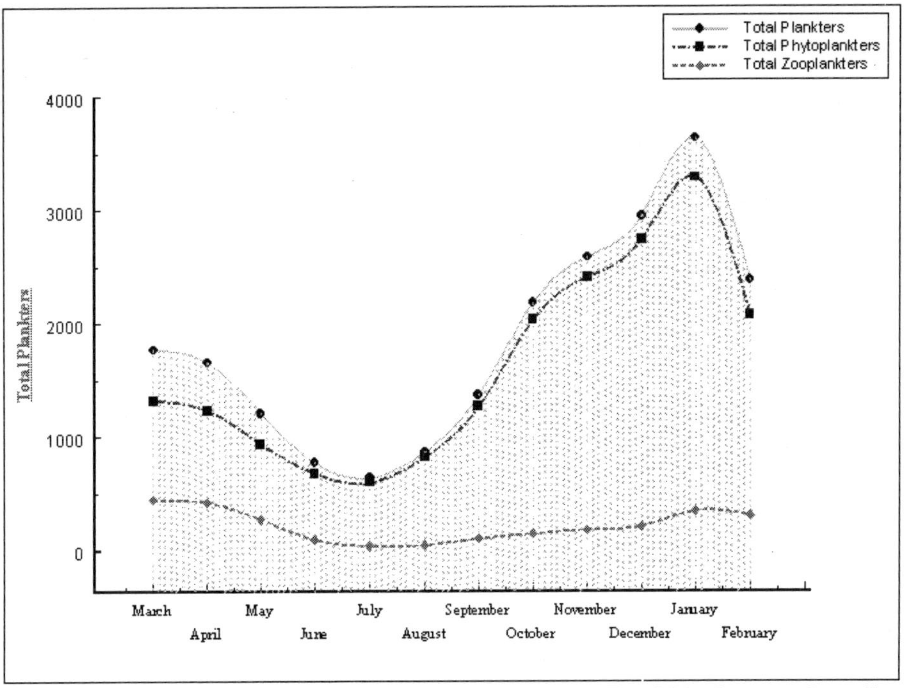

Figure 5.III.A.1: Graph Showing the Monthly Variation in Total Plankters at Sampling Site–I (2007-2008)

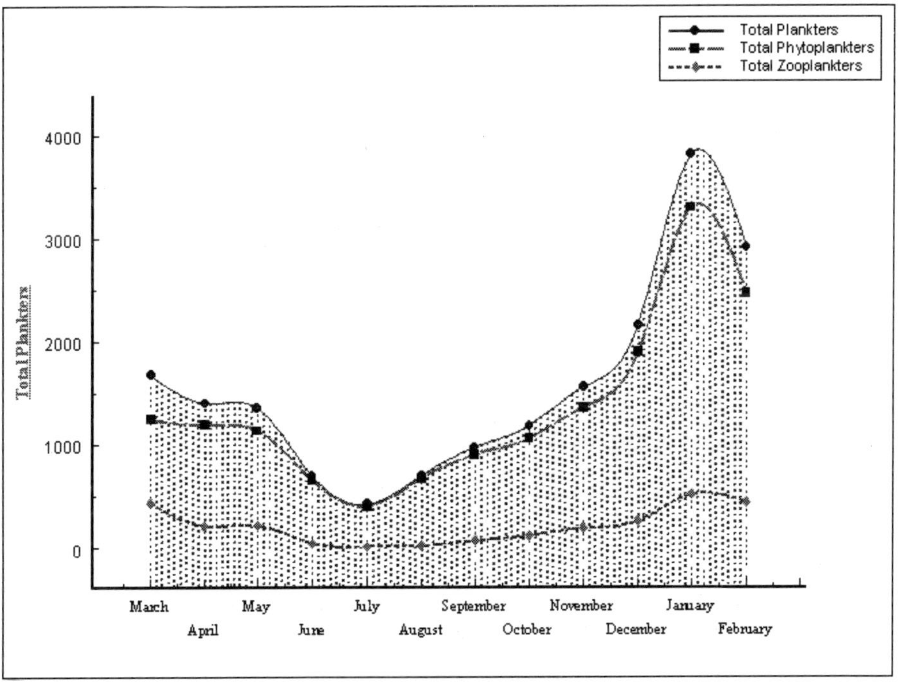

Figure 5.III.A.2: Graph Showing the Monthly Variation in Total Plankters at Sampling Site–I (2008-2009)

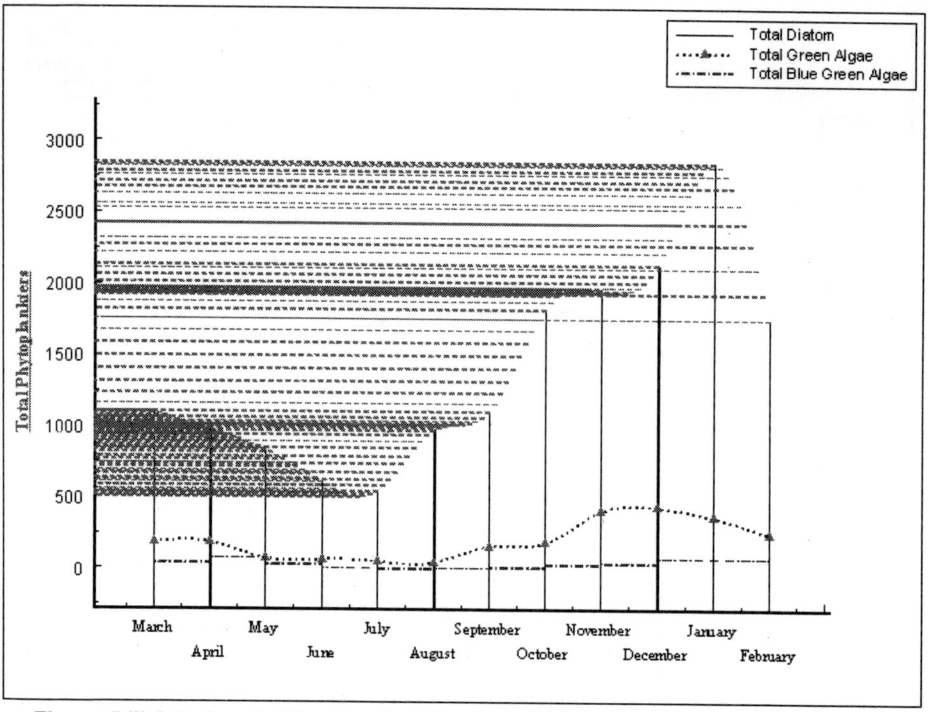

Figure 5.III.A.3: Graph Showing the Monthly Variation in Total Phytoplankters at Sampling Site–I (2007-2008)

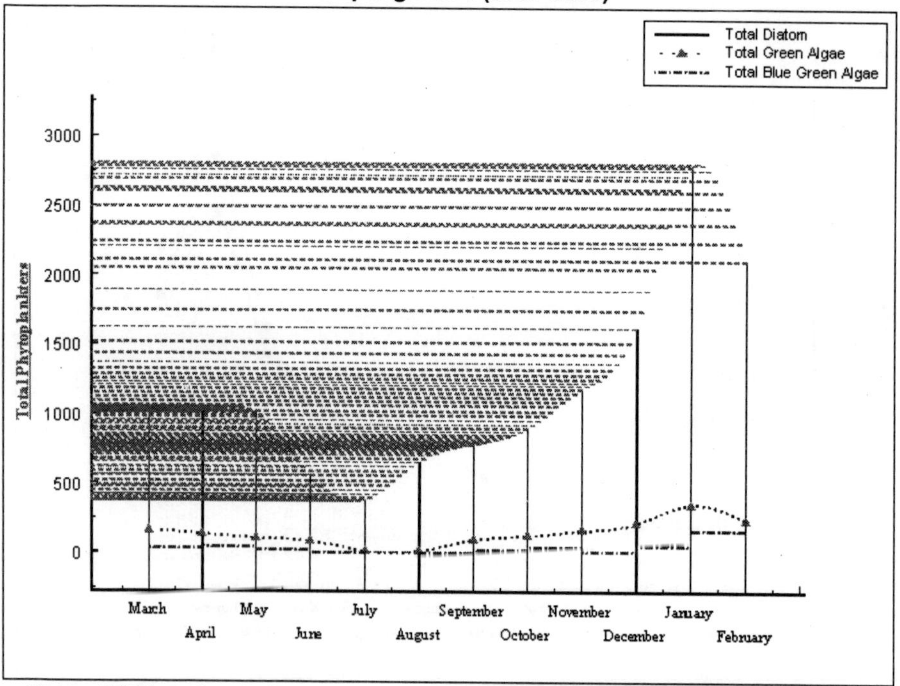

Figure 5.III.A.4: Graph Showing the Monthly Variation in Total Phytoplankters at Sampling Site–I (2008-2009)

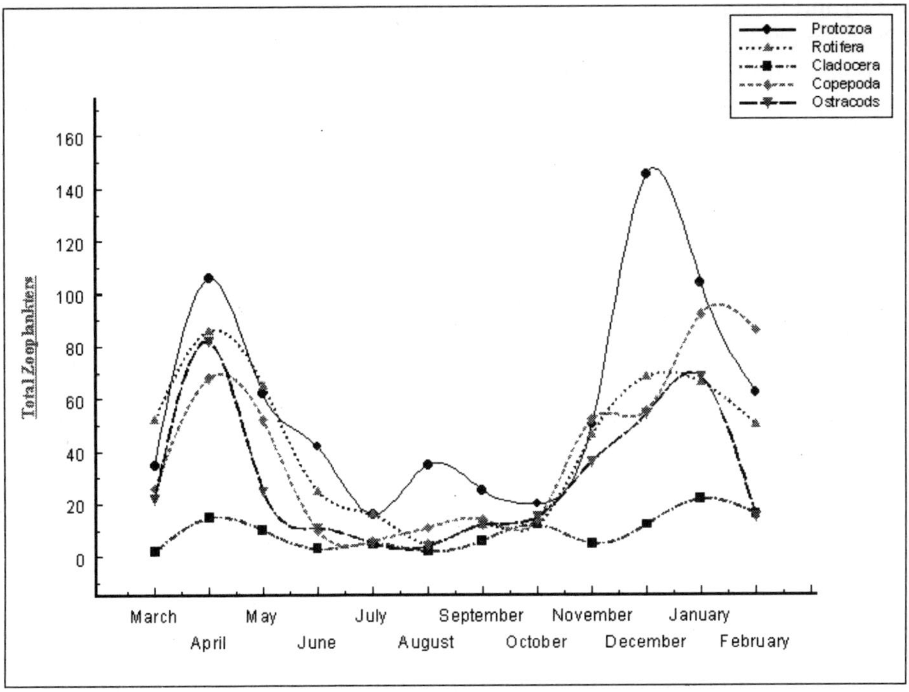

Figure 5.III.A.5: Graph Showing the Monthly Variation in Total Zooplankters at Sampling Site–I (2007-2008)

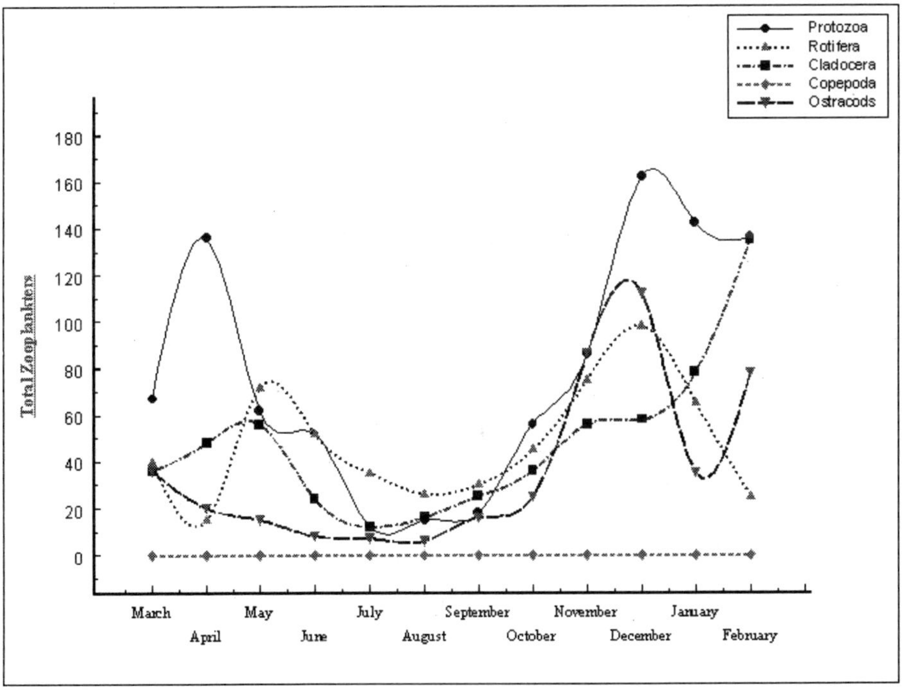

Figure 5.III.A.6: Graph Showing the Monthly Variation in Total Zooplankters at Sampling Site–I (2008-2009)

unit/l) and least peak during the month of July of 2007-2008 (2.00 unit/l). In 2007-2008 the average value of blue green algae was 36.17 unit/l ± 17.31 and in 2008-2009 it was 45.25 unit/l ± 10.94.

2. Zooplankters

Zooplankters was observed minimum in July of 2008-2009 (18.00 unit/l) and maximum in 2008-2009 for the month of January (519.00 unit/l) and the average value of 218.67 unit/l ±51.87 in 2007-2008 and in 2008-2009 the average was 213.83 unit/l±38.57.

At this site *Protozoa* was maximum in the month of January in 2008-2009 (176.00 unit/l) and minimum was found 1.00 unit/l in June month (2008-2009). *Rotifera* was found to be highest in the month of January 2008-2009 (186.00 unit/l) and lowest (5.00 unit/l) in the month of July of the same year (2008-2009). *Cladocera* was observed in their optimum peak during 2008-2009 March (112.00 unit/l) and in lowest peak during the month of August (1.00 unit/l) in the year 2007-2008. *Copepod* was maximum (106.00 unit/l) in 2007-2008 February and minimum was found in the month of July 2007-2008 (8.00 unit/l). *Ostracods* sp. was observed in their optimum peak during 2007-2008 March (106.00 unit/l) and in lowest peak during the month of August (2.00 unit/l) in the year 2007-2008.

The average value revealed that *Protozoa* was dominating in the zooplankters. In 2007-2008 (61.00 unit/l ±13.53) followed by *Rotifera* (55.65 unit/l ±12.07). *Copepod* (51.83 unit/l ±13.44), *Ostracods* sp. (38.67 unit/l ±9.61), *Cladocera* (11.33 unit/l ±3.22). In second year 2008-2009 *Rotifera* (77.75 unit/l ±11.43) found in strength followed by *Protozoa* (55.41 unit/l ± 9.86), *Ostracods* sp. (43.50 unit/l ± 11.10) and *Cladocera* (37.17 unit/l ±6.19).

Sampling Site–II

(*a*) Plankters

Quantitative analysis of plankters at Sampling site–II revealed that the total Plankters number per liter of water was observed maximum 5812.00 unit/l in the month of February in the year 2007-2008 and minimum was 636.00 unit/l in July month of the 2007-2008. The average of plankters in 2007-2008 is 2543.00 unit/l ±305.03 and in 2008-2009 is 1908.25 unit/l ±455.82 (Table 5.III.A.4 to 5.III.A.6 and Figure 5.III.A.7 to 5.III.A.12)

1. Phytoplankters

Phytoplankters was noted lowest of 604.00 unit/l in the month of July of 2007-2008 and highest of 5166.00 unit/l in February month of 2007-2008 with the average cells number 2207.33 unit/l ±329.62 in 2007-2008 and 1693.17 unit/l ±223.60 in 2008-2009.

In phytoplankters diatoms were dominated and class blue green algae was found least in both the years of study. At this sampling site diatoms was observed maximum in the month of February of 2007-2008 (4701.00 unit/l) and minimum in July month (568.00 unit/l) of 2008-2009 with average count of cells in 2007-2008 was 1728.45

Table 5.III.A.4: Quantitative Analysis of the Plankters of the Paniyala Pond During 2007-2009 at Sampling Site–II

Months	*Number of Plankters per litre of Water*				*Total Plankters (Unit/l)*	
	Phytoplankters (Unit/l)		*Zooplankters (Unit/l)*			
	2007-2008	*2008-2009*	*2007-2008*	*2008-2009*	*2007-2008*	*2008-2009*
March	4085.00	2003.00	798.00	342.00	4883.00	2345.00
April	3508.00	1235.00	557.00	331.00	4065.00	1566.00
May	1440.00	1074.00	309.00	229.00	1749.00	1303.00
June	887.00	989.00	87.00	102.00	974.00	1091.00
July	604.00	670.00	32.00	55.00	636.00	725.00
August	781.00	839.00	53.00	65.00	834.00	904.00
September	1136.00	959.00	114.00	98.00	1250.00	1057.00
October	1323.00	1957.00	164.00	149.00	1488.00	2106.00
November	1471.00	2396.00	255.00	188.00	1726.00	2584.00
December	1458.00	2971.00	387.00	260.00	1845.00	3231.00
January	4628.00	3017.00	626.00	375.00	5254.00	3392.00
February	5166.00	2205.00	646.00	390.00	5812.00	2595.00
Average±SD	2207.33±329.62	1693.17±223.60	335.67±79.10	215.33±60.13	2543.00±305.03	1908.25±455.82

SD ±: Standard Deviation.

Table 5.III.A.5: Number of Different Group Among the Phytoplankters of the Paniyala Pond During 2007-2009 at Sampling Site–II

Months	*Total Phytoplankters/ Liter of Water (Unit/l)*		*Total Diatoms/ Liter of Water (Unit/l)*		*Total Green Algae/ Liter of Water (Unit/l)*		*Total Blue Green Algae/ Liter of Water (Unit/l)*	
	2007-2008	*2008-2009*	*2007-2008*	*2008-2009*	*2007-2008*	*2008-2009*	*2007-2008*	*2008-2009*
March	4085.00	2003.00	3705.00	1681.00	311.00	240.00	68.00	82.00
April	3508.00	1235.00	3170.00	856.00	274.00	215.00	63.00	164.00
May	1440.00	1074.00	1152.00	826.00	232.00	200.00	56.00	48.00
June	887.00	989.00	805.00	868.00	75.00	115.00	6.00	6.00
July	604.00	670.00	579.00	568.00	20.00	96.00	5.00	6.00
August	781.00	839.00	739.00	732.00	35.00	98.00	7.00	8.00
September	1136.00	959.00	913.00	814.00	170.00	125.00	52.00	20.00
October	1324.00	1957.00	1075.00	1781.00	189.00	164.00	59.00	12.00
November	1471.00	2396.00	1159.00	2165.00	274.00	226.00	37.00	50.00
December	1458.00	2971.00	1015.00	2621.00	375.00	278.00	67.00	72.00
January	4628.00	3017.00	4073.00	2562.00	480.00	369.00	75.00	86.00
February	5166.00	2205.00	4701.00	1873.00	397.00	256.00	67.00	76.00
Average± SD	2207.69± 329.62	1693.72± 223.60	1728.45± 256.70	1445.58± 111.94	213.81± 54.26	198.50± 47.28	44.27± 18.66	48.75± 19.38

SD ±: Standard Deviation.

Table 5.III.A.6: Number of Different Genera Among Zooplankters in the Paniyala Pond During 2007-2009 at Sampling Site–II

Months	*Protozoa (Unit/l)*		*Rotifera (Unit/l)*		*Cladocera (Unit/l)*		*Copepod (Unit/l)*		*Ostracods (Unit/l)*	
	2007-2008	*2008-2009*	*2007-2008*	*2008-2009*	*2007-2008*	*2008-2009*	*2007-2008*	*2008-2009*	*2007-2008*	*2008-2009*
March	172.00	105.00	149.00	112.00	190.00	68.00	135.00	0.00	152.00	57.00
April	110.00	100.00	107.00	106.00	172.00	85.00	85.00	0.00	83.00	40.00
May	74.00	86.00	78.00	56.00	52.00	55.00	53.00	0.00	52.00	32.00
June	14.00	56.00	21.00	24.00	14.00	12.00	14.00	0.00	24.00	10.00
July	6.00	25.00	12.00	15.00	2.00	10.00	3.00	0.00	9.00	5.00
August	10.00	15.00	15.00	16.00	10.00	17.00	5.00	0.00	13.00	17.00
September	32.00	16.00	30.00	25.00	7.00	32.00	17.00	0.00	28.00	25.00
October	34.00	25.00	42.00	32.00	32.00	52.00	28.00	0.00	28.00	40.00
November	52.00	36.00	68.00	45.00	49.00	57.00	36.00	0.00	50.00	50.00
December	94.00	75.00	97.00	86.00	64.00	35.00	72.00	0.00	60.00	64.00
January	182.00	86.00	106.00	112.00	137.00	112.00	86.00	0.00	115.00	65.00
February	157.00	108.00	115.00	100.00	189.00	112.00	64.00	0.00	121.00	70.00
Average±SD	78.08±19.86	61.08±15.27	70.00±27.46	60.75±12.45	76.50±20.22	53.92±12.45	49.83±11.93	0.00±0.00	61.25±15.23	39.58±11.37

SD ±: Standard Deviation.

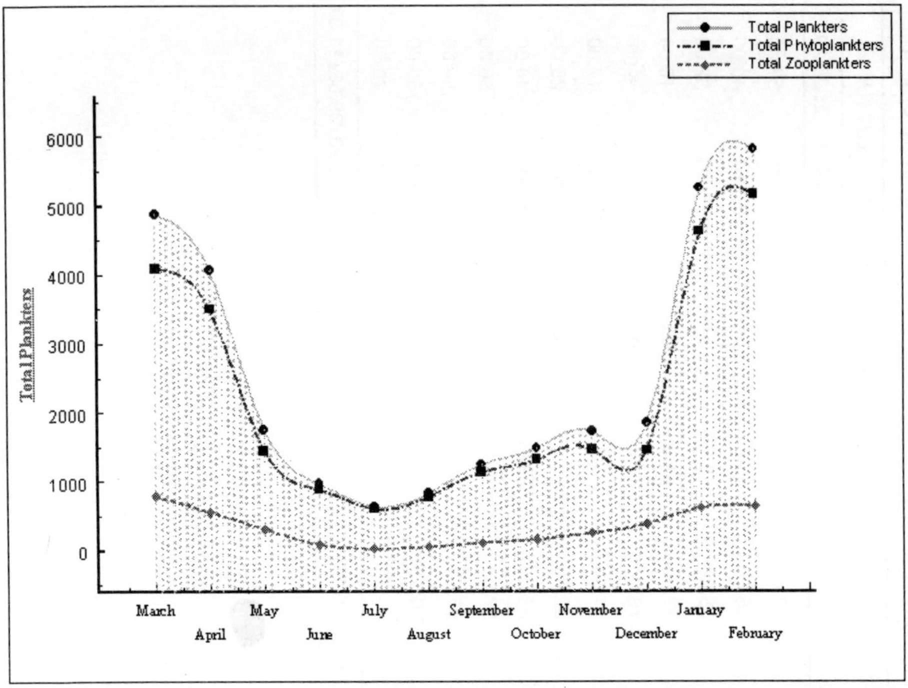

Figure 5.III.A.7: Graph Showing the Monthly Variation in Total Plankters at Sampling Site–II (2007-2008)

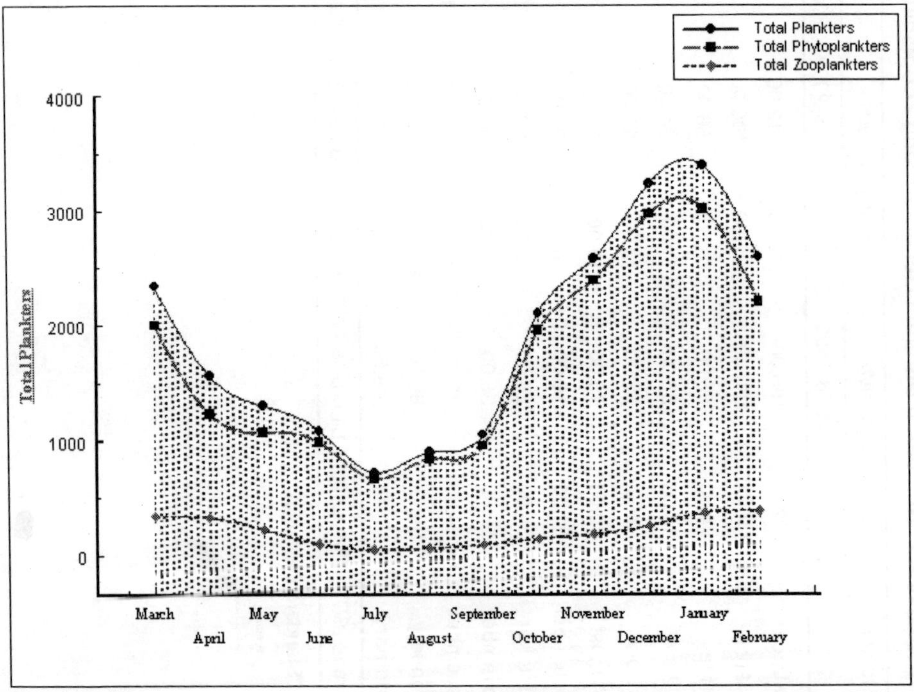

Figure 5.III.A.8: Graph Showing the Monthly Variation in Total Plankters at Sampling Site–II (2008-2009)

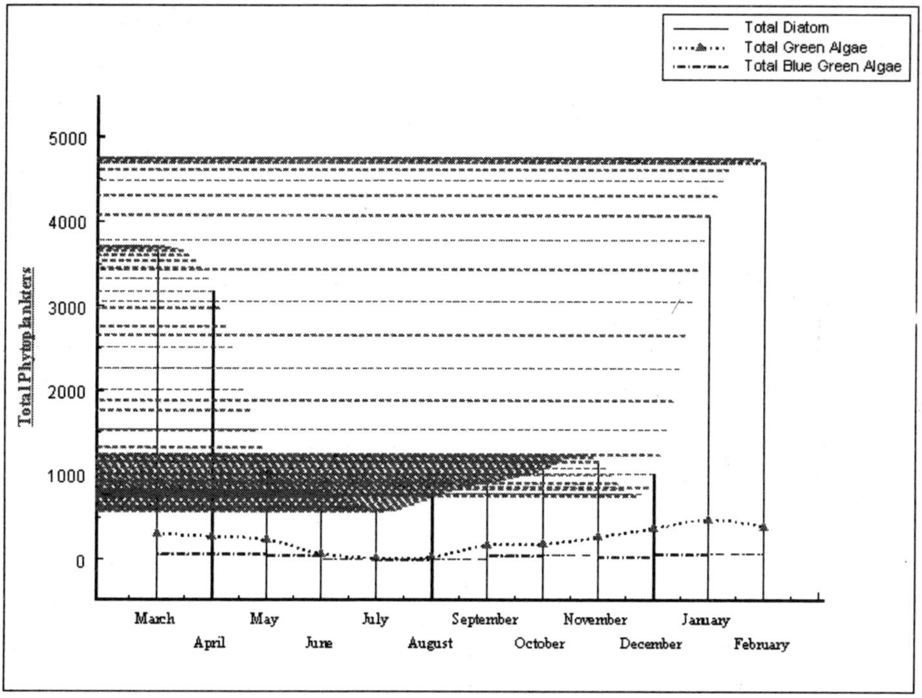

Figure 5.III.A.9: Graph Showing the Monthly Variation in Total Phytoplankters at Sampling Site–II (2007-2008)

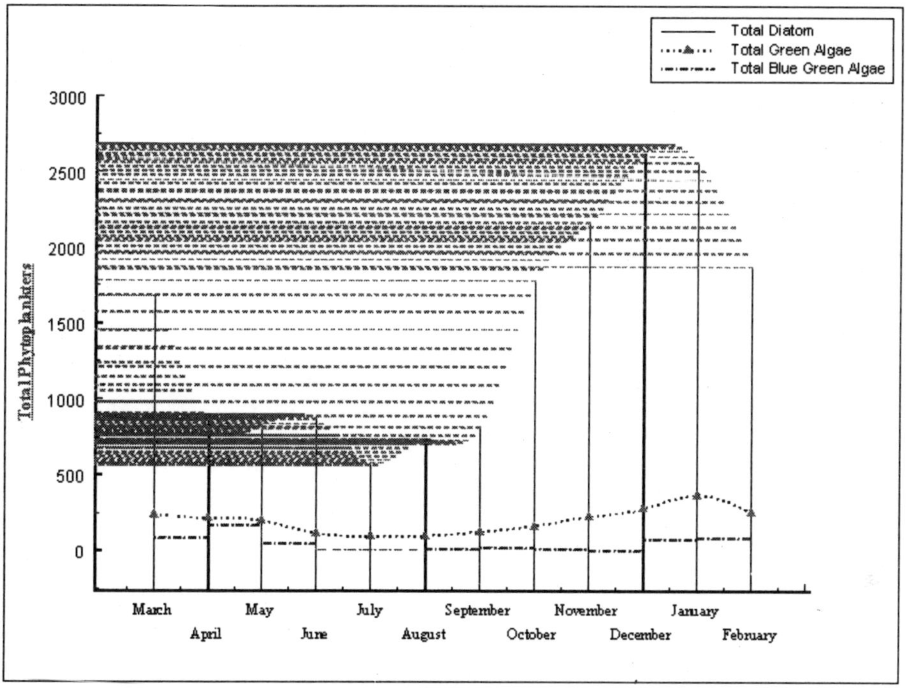

Figure 5.III.A.10: Graph Showing the Monthly Variation in Total Phytoplankters at Sampling Site–II (2008-2009)

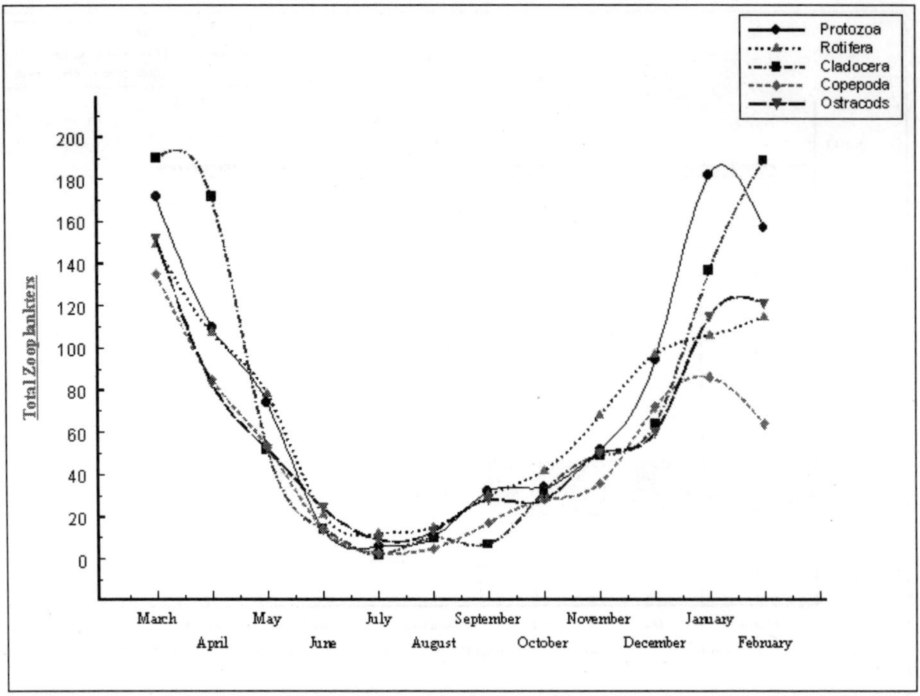

Figure 5.III.A.11: Graph Showing the Monthly Variation in Total Zooplankters at Sampling Site–II (2007-2008)

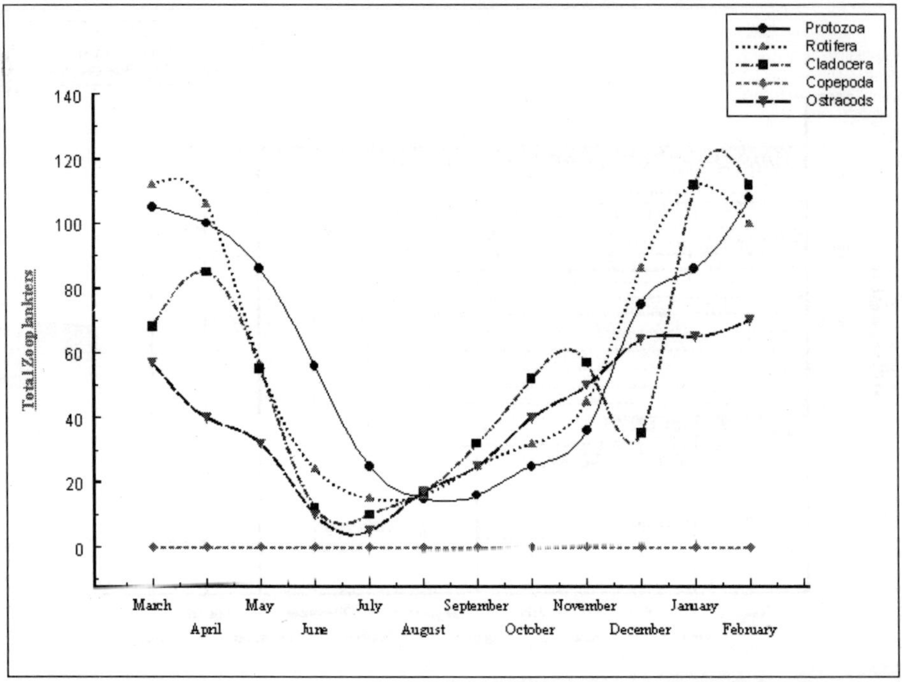

Figure 5.III.A.12: Graph Showing the Monthly Variation in Total Zooplankters at Sampling Site–II (2008-2009)

unit/l ± 256.70 and 1445.58 unit/l ± 111.94 in 2008-2009. Green algae were found to be highest in the month of January 2007-2008 (480.00 unit/l) and lowest in 2007-2008 in the month of July (20.00 unit/l). The average count of green algae in 2007-2008 was 213.81 unit/l ± 54.26 and 198.50 unit/l ± 47.28 in 2008-2009. Blue green algae showed optimum peak during the month of April in 2008-2009 (164.00 unit/l) and least peak during the month of July in 2007-2008 (5.00 unit/l) respectively. In 2007-2008 the average value of blue green algae was 44.27 unit/l ± 18.66 and in 2008-2009 it was 48.75 unit/l ±19.38.

2. Zooplankters

Zooplankters were observed minimum in July of 2007-2008 (32.00 unit/l) and maximum in 2007-2008 for the month of March (798.00 unit/l) and the average value 335.67 unit/l±79.10 in 2007-2008 and in 2008-2009 the average was 215.33 unit/l±60.13.

At this site *Protozoa* was maximum in the month of January in 2007-2008 (182.00 unit/l) and minimum was found 6.00 unit/l in July month (2007-2008). *Rotifera* was found to be highest in the month of March 2007-2008 (149.00 unit/l) and lowest (12.00 unit/l) in the month of July of the same year (2007-2008). *Cladocera* was observed in their optimum peak during 2007-2008 March (190.00 unit/l) and in lowest peak during the month of July (2.00 unit/l) in the year 2007-2008.*Copepod* was maximum (135.00 unit/l) in 2007-2008 March and minimum was found in the month of July 2007-2008 (3.00 unit/l). *Ostracods* sp. was observed in their optimum peak during 2007-2008 March (152.00 unit/l) and in lowest peak during the month of July (5.00 unit/l) in the year 2008-2009.

The average value revealed that *Protozoa* was dominating in the zooplankters, 2007-2008 (78.08 unit/l ±19.86) followed by *Cladocera* (76.50 unit/l±20.22). *Rotifera* (70 unit/l ± 27.46), *Ostracods* (61.25 unit/l ± 15.23) *Copepod* (49.83 unit/l±11.93). In second year 2008-2009 *Protozoa* (61.08 unit/l ± 15.27) found in strength followed by *Rotifera* (60.75 ± 12.45) and *Cladocera* (53.92 unit/l ± 12.45) and *Ostracods* (39.58 unit/l ± 11.37).

Sampling Site–III

(a) Plankters

Quantitative analysis of plankters at Sampling site–III revealed that the total Plankters number per liter of water was observed maximum 4847.00 unit/l in the month of February in the year 2007-2008 and minimum was 699.00 unit/l in July month of the 2007-2008. The average value of plankters in 2007-2008 is 2318.17 ± 343.17 and in 2008-2009 is 1761.75 ± 190.63 (Table 5.III.A.7 to 5.III.A.9 and Figure 5.III.A.13 to 5.III.A.18)

1. Phytoplankters

Phytoplankters was noted lowest of 659.00 unit/l in the month of July of 2007-2008 and highest of 4486.00 unit/l in February month of 2007-2008 with the average cells number 2096.42 unit/l ± 257.14 in 2007-2008 and 1539.83 unit/l ± 144.61 in 2008-2009.

Table 5.III.A.7: Quantitative Analysis of the Plankters of the Paniyala Pond During 2007-2009 at Sampling Site–III

Months	*Number of Plankters per litre of Water*				*Total Plankters (Unit/l)*	
	Phytoplankters (Unit/l)		*Zooplankters (Unit/l)*			
	2007-2008	*2008-2009*	*2007-2008*	*2008-2009*	*2007-2008*	*2008-2009*
March	3319.00	1187.00	458.00	253.00	3777.00	1440.00
April	2472.00	1084.00	318.00	275.00	2790.00	1359.00
May	2845.00	974.00	183.00	128.00	3028.00	1102.00
June	1121.00	933.00	92.00	61.00	1213.00	994.00
July	659.00	735.00	40.00	42.00	699.00	777.00
August	854.00	729.00	58.00	69.00	912.00	798.00
September	1025.00	1352.00	134.00	123.00	1159.00	1475.00
October	1237.00	1836.00	182.00	290.00	1419.00	2126.00
November	1225.00	1984.00	240.00	225.00	1465.00	2209.00
December	2471.00	2603.00	273.00	267.00	2744.00	2870.00
January	3443.00	2635.00	322.00	580.00	3765.00	3215.00
February	4486.00	2423.00	361.00	350.00	4847.00	2773.00
Average±SD	2096.42±257.14	1539.83±144.61	221.76±91.31	213.55±52.21	2318.17±343.17	1761.75±190.63

SD ±: Standard Deviation.

Table 5.III.A.8: Number of Different Group Among the Phytoplankters of the Paniyala Pond During 2007-2009 at Sampling Site–III

Months	*Total Phytoplankters/ Liter of Water (Unit/l)*		*Total Diatoms/ Liter of Water (Unit/l)*		*Total Green Algae/ Liter of Water (Unit/l)*		*Total Blue Green Algae/ Liter of Water (Unit/l)*	
	2007-2008	*2008-2009*	*2007-2008*	*2008-2009*	*2007-2008*	*2008-2009*	*2007-2008*	*2008-2009*
March	3319.00	1187.00	3100.00	987.00	178.00	200.00	40.00	77.00
April	2472.00	1084.00	2274.00	968.00	146.00	116.00	52.00	28.00
May	2845.00	974.00	2692.00	868.00	126.00	106.00	26.00	26.00
June	1121.00	933.00	987.00	835.00	75.00	98.00	58.00	6.00
July	659.00	735.00	639.00	715.00	14.00	20.00	4.00	5.00
August	854.00	729.00	832.00	684.00	15.00	37.00	6.00	8.00
September	1025.00	1352.00	936.00	1210.00	12.00	127.00	16.00	15.00
October	1237.00	1836.00	1020.00	1652.00	186.00	168.00	30.00	16.00
November	1225.00	1984.00	960.00	1721.00	265.00	262.00	9.00	10.00
December	2471.00	2603.00	2172.00	2286.00	326.00	315.00	84.00	72.00
January	3443.00	2635.00	3032.00	2267.00	326.00	368.00	85.00	92.00
February	4486.00	2423.00	4152.00	2181.00	278.00	242.00	56.00	67.00
Average± SD	2096.42± 257.14	1539.83± 144.61	1899.71± 228.06	1364.50± 121.17	162.26± 20.28	171.58± 11.25	38.85± 8.80	35.17 ± 12.19

SD ±: Standard Deviation.

Table 5.III.A.9: Number of Genera Among Zooplankters in the Paniyala Pond During 2007-2009 at Sampling Site–III

Months	*Protozoa (Unit/l)*		*Rotifera (Unit/l)*		*Cladocera (Unit/l)*		*Ostracods (Unit/l)*	
	2007-2008	*2008-2009*	*2007-2008*	*2008-2009*	*2007-2008*	*2008-2009*	*2007-2008*	*2008-2009*
March	114.00	25.00	150. 00	125.00	107.00	46.00	87.00	57.00
April	112.00	17.00	16.00	117.00	125.00	96.00	65.00	45.00
May	72.00	12.00	64.00	52.00	32.00	52.00	15.00	12.00
June	30.00	11.00	35.00	25.00	15.00	16.00	12.00	9.00
July	10.00	10.00	11.00	15.00	12.00	11.00	7.00	6.00
August	14.00	15.00	14.00	25.00	15.00	15.00	15.00	14.00
September	32.00	28.00	25.00	36.00	37.00	24.00	40.00	35.00
October	40.00	45.00	30.00	38.00	56.00	62.00	56.00	45.00
November	45.00	62.00	48.00	40.00	84.00	67.00	63.00	56.00
December	85.00	75.00	85.00	72.00	28.00	35.00	75.00	85.00
January	149.00	96.00	21.00	196.00	112.00	116.00	40.00	172.00
February	119.00	32.00	15.00	175.00	137.00	58.00	90.00	85.00
Average±Sd	68.50±21.35	35.67±8.61	42.83±26.03	76.33±16.32	63.33±16.55	49.83 ±10.53	47.08±29.20	51.75±11.40

SD ±: Standard Deviation.

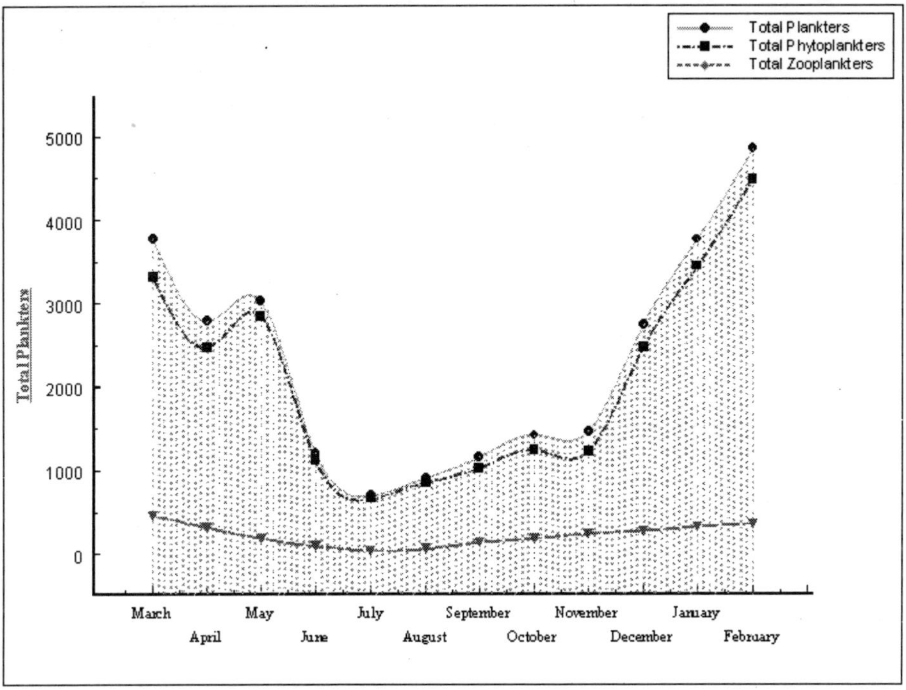

Figure 5.III.A.13: Graph Showing the Monthly Variation in Total Plankters at Sampling Site–III (2007-2008)

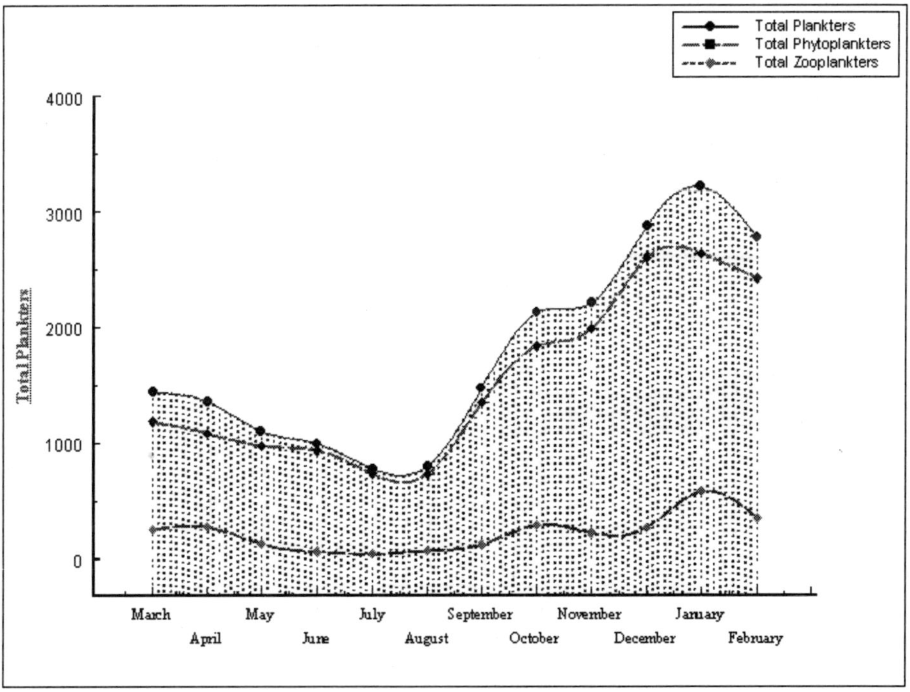

Figure 5.III.A.14: Graph Showing the Monthly Variation in Total Plankters at Sampling Site–III (2008-2009)

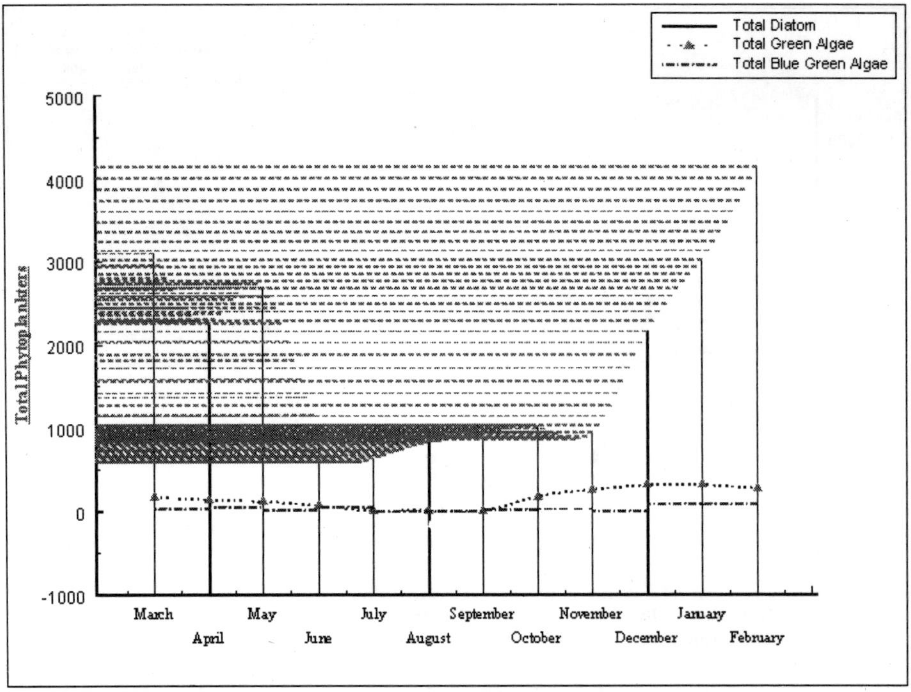

Figure 5.III.A.15: Graph Showing the Monthly Variation in Total Phytoplankters at Sampling Site–III (2007-2008)

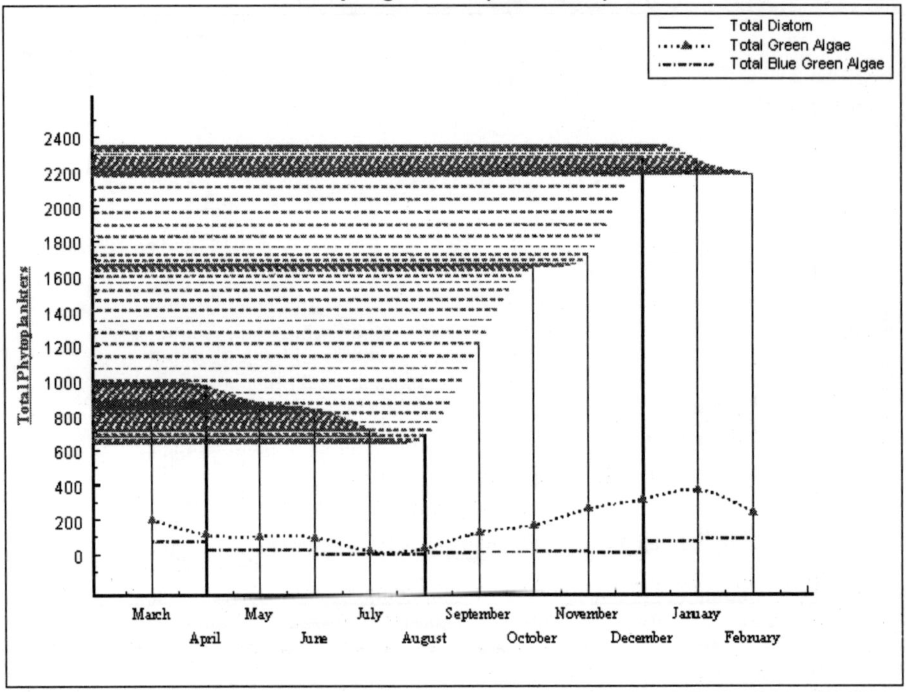

Figure 5.III.A.16: Graph Showing the Monthly Variation in Total Phytoplankters at Sampling Site–III (2008-2009)

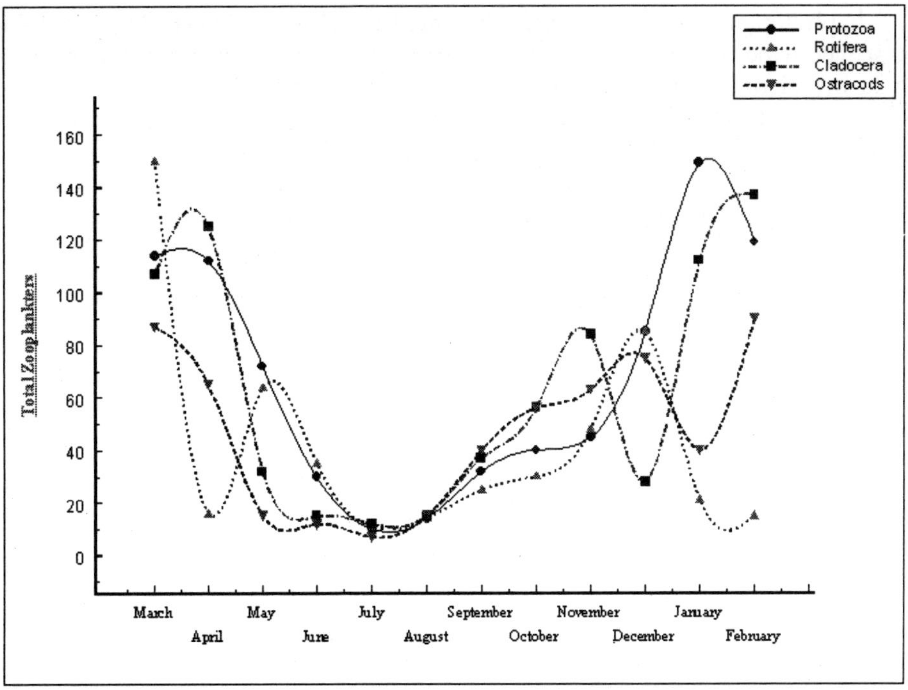

Figure 5.III.A.17: Graph Showing the Monthly Variation in Total Zooplankters at Sampling Site–III (2007-2008)

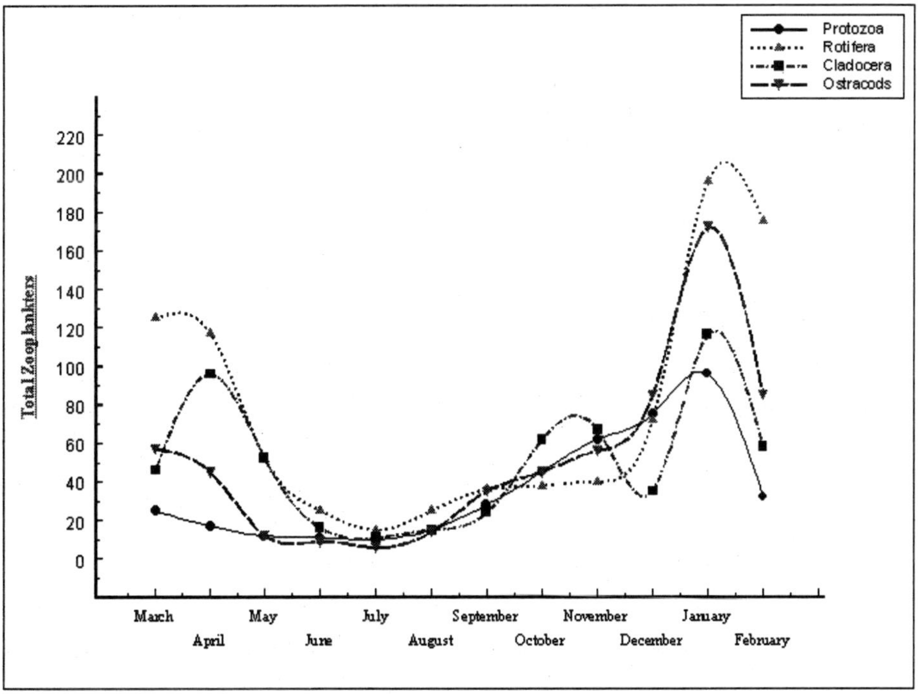

Figure 5.III.A.18: Graph Showing the Monthly Variation in Total Zooplankters at Sampling Site–III (2008-2009)

In phytoplankters diatoms was dominated and class blue green algae was found least in both the years of study. At this sampling site diatoms was observed maximum in the month of February of 2007-2008 (4152.00 unit/l) and minimum in July month (639.00 unit/l) of 2007-2008 with average count of cells in 2007-2008 was 1899.71 unit/l± 228.06 and 1364.50 unit/l± 121.17 in 2008-2009. Green algae were found to be highest in the month of January 2008-2009 (368.00 unit/l) and lowest in 2007-2008 in the month of September (12.00 unit/l). The average count of green algae in 2007-2008 was 162.26 unit/l ± 20.28 and 171.58 unit/l± 11.25 in 2008-2009. Blue green algae showed optimum peak during the month of January in 2008-2009 (92.00 unit/l) and least peak during the month of July of 2007-2008 (4.00 unit/l). In 2007-2008 the average value of Blue green algae was 38.85 unit/l ± 8.80 and in 2008-2009 it was 35.17 unit/l ± 12.19.

2. Zooplankters

Zooplankters was observed minimum in July of 2007-2008 (40.00 unit/l) and maximum in 2008-2009 for the month of January (580.00 unit/l) and the average value of 221.76 unit/l ± 91.31 in 2007-2008 and in 2008-2009 the average was 213.55 unit/l ± 52.21.

At this site *Protozoa* was maximum in the month of January in 2007-2008 (149.00 unit/l) and minimum was found 10.00 unit/l in July month in 2007-2008 and 2008-2009. *Rotifera* was found to be highest in the month of January 2008-2009 (196.00 unit/l) and lowest (11.00 unit/l) in the month of July of the year 2007-2008. *Cladocera* was observed in their optimum peak during 2007-2008 February (137.00 unit/l) and in lowest peak during the month of July (11.00 unit/l) in the year 2008-2009. *Ostracods* sp. was observed in their optimum peak during 2008-2009 January (172.00 unit/l) and in lowest peak during the month of July (6.00 unit/l) in the year 2008-2009.

The average value revealed that *Protozoa* was dominating in the zooplankters, 2007-2008 (68.50 unit/l ± 21.35) followed by *Cladocera* (63.33 unit/l ± 16.55), *Ostracods* sp. (47.08 unit/l ± 29.20), *Rotifera* (42.83 unit/l ± 26.03). In second year 2008-2009 *Rotifera* (76.33 unit/l ± 16.32) found in strength followed by *Ostracods* sp. (51.75 unit/ l ± 11.40), *Cladocera* (49.83 unit/l ± 10.53) and *Protozoa* (35.67 unit/l ± 8.61).

Sampling Site–IV

(a) Plankters

Quantitative analysis of plankters at Sampling site–IV revealed that the total Plankters number per liter of water was observed maximum 4357.00 unit/l in the month of January in the year 2008-2009 and minimum was 565.00 unit/l in July month of the 2007-2008. The average value of plankters in 2007-2008 is 1558.92 ± 156.29 and 2008-2009 is 2211.25 ± 231.41 (Table 5.III.A.10 to 5.III.A.12 and Figure 5.III.A.19 to 5.III.A.24)

1. Phytoplankters

Phytoplankters was noted lowest of 517.00 unit/l in the month of July of 2007-2008 and highest of 4129.00 unit/l in January month of 2008-2009 with the average

Table 5.III.A.10: Quantitative Analysis of the Plankters of the Paniyala Pond During 2007-2009 at Sampling Site–IV

Months	*Number of Plankters per litre of Water*				*Total Plankters (Unit/l)*	
	Phytoplankters (Unit/l)		*Zooplankters (Unit/l)*			
	2007-2008	*2008-2009*	*2007-2008*	*2008-2009*	*2007-2008*	*2008-2009*
March	1282.00	1352.00	137.00	221.00	1419.00	1573.00
April	1259.00	2449.00	357.00	389.00	1616.00	2838.00
May	1257.00	1979.00	214.00	205.00	1471.00	2184.00
June	728.00	999.00	91.00	136.00	819.00	1135.00
July	517.00	635.00	48.00	66.00	565.00	701.00
August	631.00	821.00	57.00	63.00	688.00	884.00
September	688.00	1112.00	69.00	89.00	757.00	1201.00
October	693.00	1073.00	73.00	162.00	766.00	1235.00
November	1471.00	2941.00	189.00	303.00	1660.00	3244.00
December	2431.00	3560.00	334.00	397.00	2765.00	3957.00
January	3169.00	4129.00	352.00	228.00	3521.00	4357.00
February	2369.00	2939.00	229.00	287.00	2660.00	3226.00
Average±SD	1374.58±117.69	1999.08±191.35	179.17±87.93	212.17±42.00	1558.92±156.29	2211.25±231.41S

SD ±: Standard Deviation.

Table 5.III.A.11: Number of Different Group Among the Phytoplankters of the Paniyala Pond During 2007-2009 at Sampling Site–IV

Months	*Total Phytoplankters/ Liter of Water (Unit/l)*		*Total Diatoms/ Liter of Water (Unit/l)*		*Total Green Algae/ Liter of Water (Unit/l)*		*Total Blue Green Algae/ Liter of Water (Unit/l)*	
	2007-2008	*2008-2009*	*2007-2008*	*2008-2009*	*2007-2008*	*2008-2009*	*2007-2008*	*2008-2009*
March	1282.00	1352.00	1065.00	1075.00	212.00	252.00	5.00	25.00
April	1259.00	2449.00	1108.00	2108.00	102.00	282.00	48.00	58.00
May	1257.00	1979.00	1161.00	1861.00	65.00	85.00	30.00	32.00
June	728.00	999.00	656.00	856.00	68.00	98.00	4.00	45.00
July	517.00	635.00	462.00	562.00	54.00	64.00	1.00	9.00
August	631.00	821.00	586.00	786.00	42.00	32.00	3.00	3.00
September	688.00	1112.00	615.00	985.00	65.00	85.00	8.00	42.00
October	693.00	1073.00	565.00	865.00	105.00	125.00	22.00	82.00
November	1471.00	2941.00	1214.00	2614.00	202.00	262.00	55.00	65.00
December	2431.00	3560.00	2101.00	3100.00	268.00	378.00	62.00	82.00
January	3169.00	4129.00	3021.00	3721.00	102.00	362.00	46.00	46.00
February	2369.00	2939.00	2205.00	2765.00	106.00	146.00	58.00	28.00
Average± SD	1375.10± 117.69	1999.68± 191.35	1229.92± 94.16	1775.10± 152.92	115.92± 13.80	181.26± 26.16	28.50± 9.73	43.32± 12.27

SD ±: Standard Deviation.

Table 5.III.A.12 Number of Different Genera among Zooplankters in the Paniyala Pond During 2007-2009 at Sampling Site–IV

Months	*Protozoa (Unit/l)*		*Rotifera (Unit/l)*		*Cladocera (Unit/l)*		*Copepod (Unit/l)*		*Ostracods (Unit/l)*	
	2007-2008	*2008-2009*	*2007-2008*	*2008-2009*	*2007-2008*	*2008-2009*	*2007-2008*	*2008-2009*	*2007-2008*	*2008-2009*
March	35.00	67.00	52.00	40.00	2.00	36.00	26.00	0.00	220.00	360.00
April	106.00	136.00	86.00	15.00	15.00	48.00	68.00	0.00	820.00	200.00
May	62.00	62.00	65.00	72.00	10.00	56.00	52.00	0.00	250.00	150.00
June	42.00	52.00	25.00	52.00	3.00	24.00	10.00	0.00	110.00	80.00
July	16.00	12.00	16.00	35.00	5.00	12.00	6.00	0.00	50.00	70.00
August	35.00	15.00	5.00	26.00	2.00	16.00	11.00	0.00	40.00	60.00
September	25.00	18.00	12.00	30.00	6.00	25.00	14.00	0.00	120.00	160.00
October	20.00	56.00	12.00	45.00	12.00	36.00	14.00	0.00	150.00	250.00
November	50.00	86.00	46.00	75.00	5.00	56.00	52.00	0.00	360.00	860.00
December	145.00	162.00	68.00	98.00	12.00	58.00	55.00	0.00	540.00	1120.00
January	104.00	142.00	66.00	65.00	22.00	78.00	92.00	0.00	680.00	350.00
February	62.00	136.00	50.00	25.00	16.00	135.00	86.00	0.00	150.00	780.00
Average±SD	58.50±11.77	78.67±14.51	41.92±9.22	48.17±7.51	9.17±1.95	48.33±8.08	40.50±10.65	0.00±0.00	29.08±7.05	37.00±11.89

SD ±: Standard Deviation.

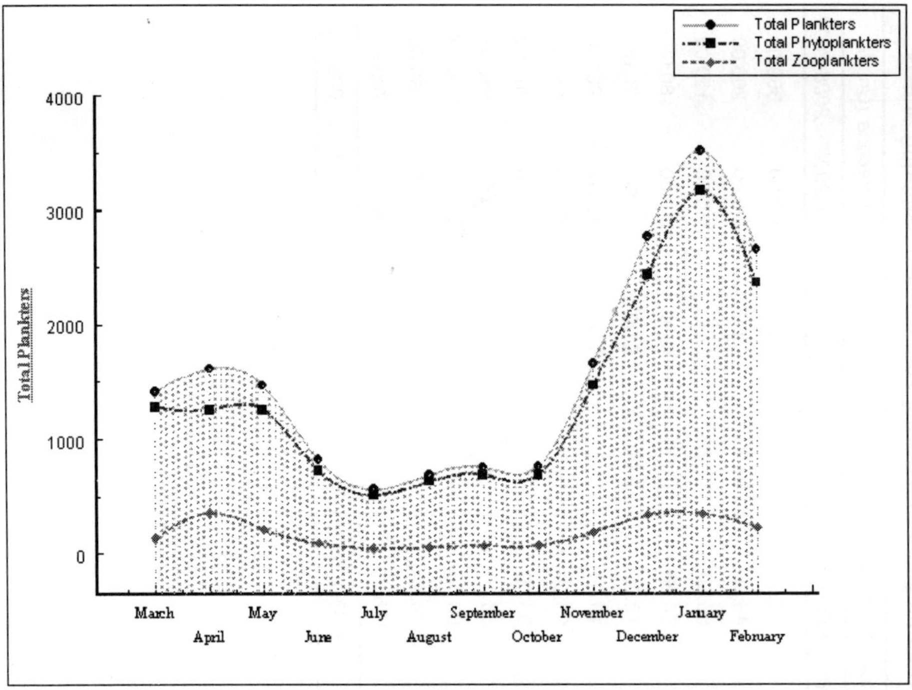

Figure 5.III.A.19: Graph Showing the Monthly Variation in Total Plankters at Sampling Site–IV (2007-2008)

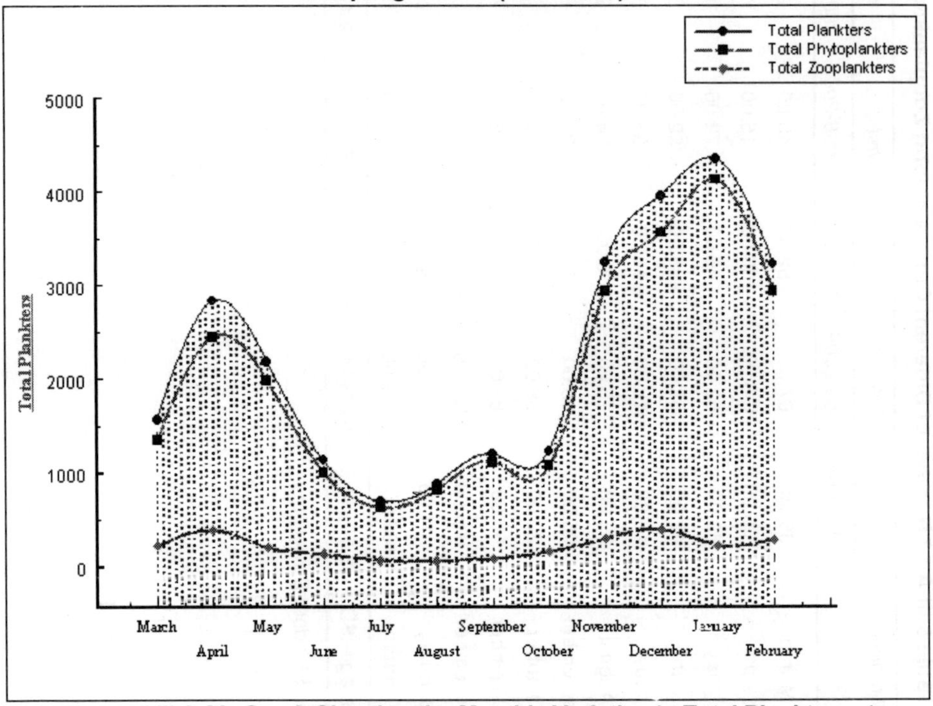

Figure 5.III.A.20: Graph Showing the Monthly Variation in Total Plankters at Sampling Site–IV (2008-2009)

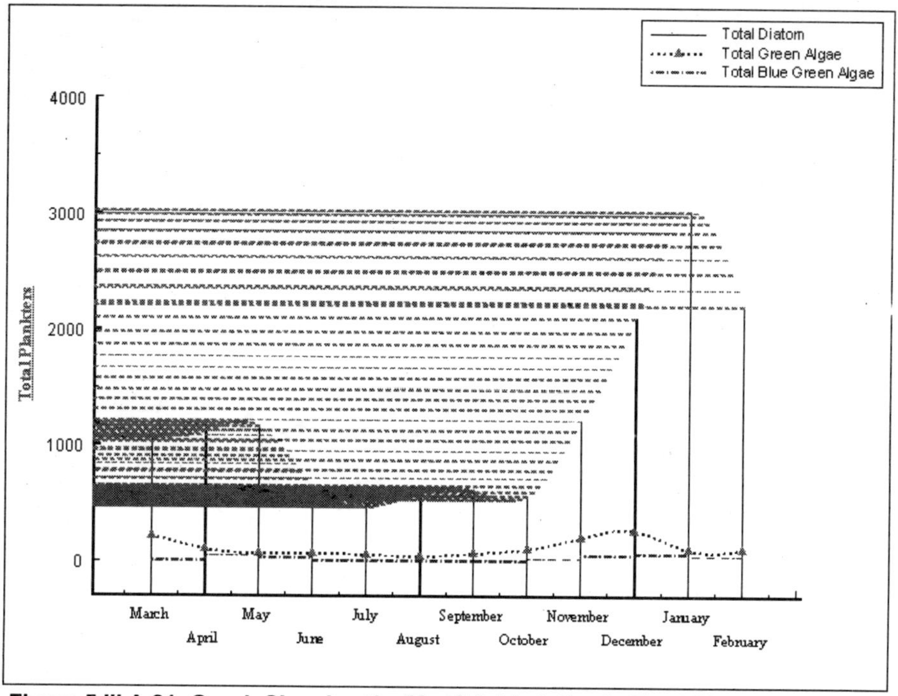

Figure 5.III.A.21: Graph Showing the Monthly Variation in Total Phytoplankters at Sampling Site–IV (2007-2008)

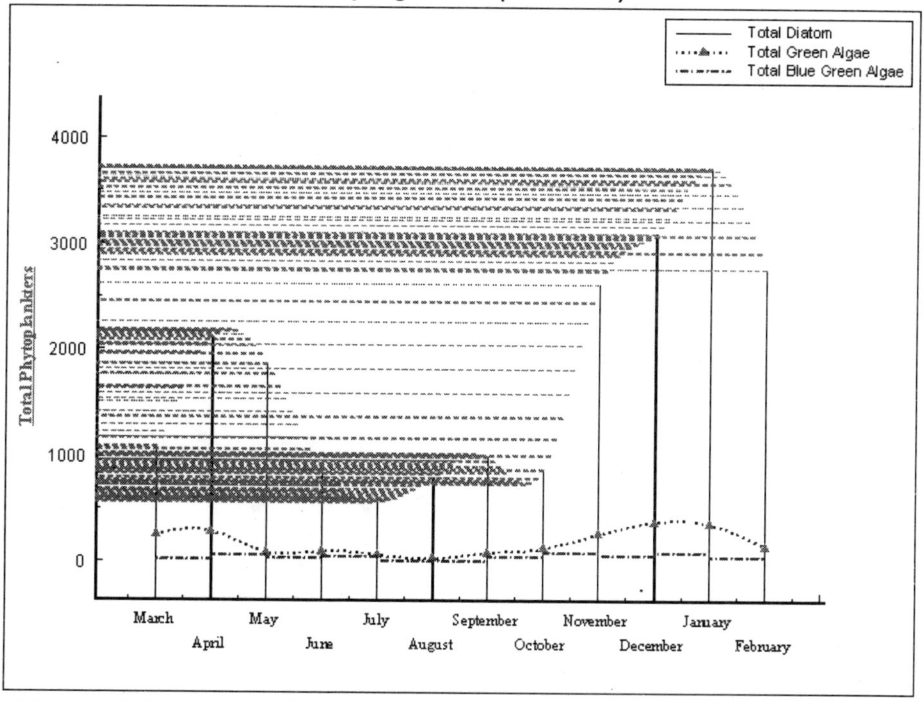

Figure 5.III.A.22: Graph Showing the Monthly Variation in Total Phytoplankters at Sampling Site–IV (2008-2009)

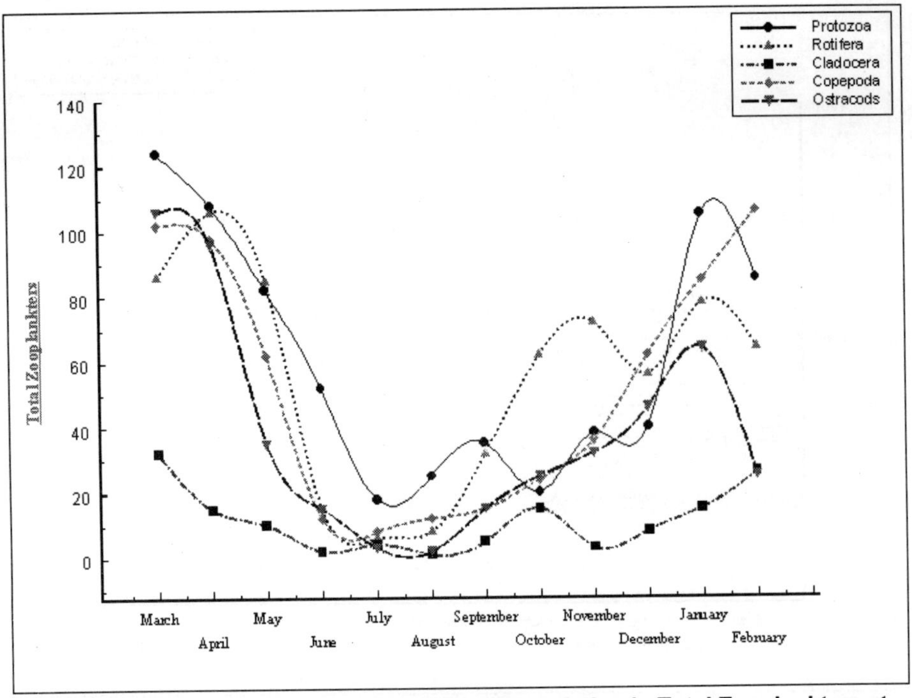

Figure 5.III.A.23: Graph Showing the Monthly Variation in Total Zooplankters at Sampling Site–IV (2007-2008)

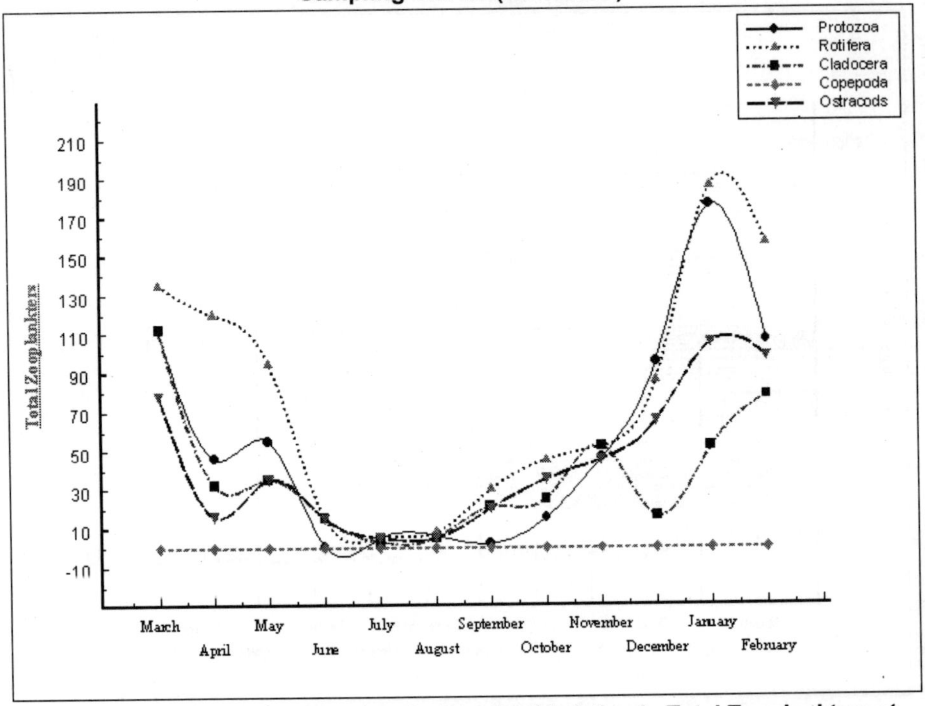

Figure 5.III.A.24: Graph Showing the Monthly Variation in Total Zooplankters at Sampling Site–IV (2008-2009)

cells number 1374.58 unit/l ± 117.69 in 2007-2008 and 1999.08 unit/l ± 191.35 in 2008-2009.

In phytoplankters diatoms was dominated and class blue green algae was found least in both the years of study. At this sampling site diatoms was observed maximum in the month of January of 2008-2009 (3721.00 unit/l) and minimum in July month (462.00 unit/l) of 2007-2008 with average count of cells in 2007-2008 was 1229.92 unit/l ±94.16 and 1775.10 unit/l ±152.92 in 2008-2009. Green algae were found to be highest in the month of December 2008-2009 (378.00 unit/l) and lowest in 2008-2009 in the month of August (32.00 unit/l). The average count of Green algae in 2007-2008 was 115.92 unit/l ± 13.80 and 181.26 unit/l± 26.16 in 2008-2009. Blue green algae showed optimum peak during the months of October and December in 2008-2009 (82.00 unit/l) and least peak during the month of July of 2007-2008 (1.00 unit/l). In 2007-2008 the average value of Blue green algae was 28.50 unit/l ± 9.73 and in 2008-2009 it was 43.32 unit/l ± 12.27.

2. Zooplankters

Zooplankters was observed minimum in July of 2007-2008 (48.00 unit/l) and maximum in 2008-2009 for the month of December (397.00 unit/l). And the average value of 179.17 unit/l ± 87.93 in 2007-2008 and in 2008-2009 the average was 212.17 unit/l ± 42.00.

At this site *Protozoa* was maximum in the month of December in 2008-2009 (162.00 unit/l) and minimum was found 12.00 unit/l in July month (2008-2009). *Rotifera* was found to be highest in the month of December 2008-2009 (98.00 unit/l) and lowest (5.00 unit/l) in the month of August of the year 2007-2008. *Cladocera* was observed in their optimum peak during 2008-2009 February (135.00 unit/l) and in lowest peak during the months of March and August (2.00 unit/l) in the year 2007-2008. *Copepod* was maximum (92.00 unit/l) in 2007-2008 January and minimum was found in the month of July 2007-2008 (6.00 unit/l). *Ostracods sp.* was observed in their optimum peak during 2008-2009 December (112.00 unit/l) and in lowest peak during the month of August (4.00 unit/l) in the year 2007-2008.

The average value revealed that *Protozoa* was dominating in the zooplankters, 2007-2008 (58.50 unit/l ± 11.77) followed by *Rotifera* (41.92 unit/l ± 9.22), *Copepod* (40.50 unit/l ± 10.60), *Ostracods* sp. (29.08 unit/l ± 7.05), *Cladocera* (9.17 unit/l ± 1.95). In second year 2008-2009 *Protozoa* (78.67 unit/l ± 14.51) found in strength followed by *Cladocera* (48.33 unit/l ± 8.08), *Rotifer* (48.17 ± 7.51), and *Ostracods sp.* (37.00 unit/l ± 11.89).

Fishes

Paniyala Fish Pond is a one of the major pond of Fisheries Department in Paniyala Village, Roorkee. During the study of Paniyal fish Pond number of fishes were observed. Some of them are culturable fishes and other species of weed fish species. Below a list of fishes were given with there common names and family.

Sl.No.	Name (Generic -Species)	Common Names	Family
1.	*Catla-catla*	Catla	Cyprinidae
2.	*Labeo-rohita*	Rohu	Cyprinidae
3.	*Cirrhinus-mrigala*	Mrigala	Cyprinidae
4.	*Hypophthalmichthys-molitrix*	Silver carp	Cyprinidae
5.	*Cyprinus-carpio*	China carp	Cyprinidae
6.	*Channa-stratius*	Dolly	Ophiocephalidae
7.	*Notopterus-notopterus*	Dhuta	Notopteridae
8.	*Chanda-nama*	Chandla	Centropomidae

B. Aquatic Flora

Aquatic plants play a key role in biogeochemical cycles and food webs of living things in the lentic water bodies. On the other hand aquatic plants are primary producers of food in aquatic ecosystem, and further the absence of aquatic flora may also lead to the absence of fauna. Distribution of macrophytic plants and abundance also indicate the water quality. The hydrophytic flora of this pond shown an interesting distribution. The foregoing account of the floristic composition of hydrophytes of Paniyala Pond showed that the aquatic and semi-aquatic plants includes pteriodophytes, dicotyledons and monocotyledons.

A variety of aquatic weeds (floating, submerged as well as emergent) are present in the catchment area of the stagnant part of the pond. In the present study, altogether 14 macrophytes species belonging to 13 families were recorded. The species were grouped under different categories *viz.*, free floating, rooted floating, submerged and emergent.

Sampling Site–I and Sampling Site–IV showed very less amount of free floating, submerged and rooted floating vegetation with few exceptions. Most of the rooted emergent macrophytes occupied the space near Sampling Site – II and Sampling Site – III. Eicchornia was the dominant free floating hydrophyte throughout the period of investigation. (Table 5.III.B.1)

C. Microbiological Characteristics

(*i*) MPN (Total Coliform Bacteria)

Sampling Site–I

The MPN/100ml of the pond at this sampling site-I was recorded minimum 07 MPN/ 100 ml in the month of January in the year 2007-2008 and maximum of 176 MPN/ 100 ml in the month of September in the year 2008-2009. The average value in the year 2007-2008 was 69.83 MPN/ 100 ml ± 56.99 and 77.25 MPN/ 100 ml ± 62.07 in the year 2008-2009. (Table 5.III.C.i.1 and Figure 5.III.C.i.1)

Table 5.III.B.1: Aquatic Flora of Paniyala Fish Pond at all the Sampling Site–I in 2007-2009

Sl.No.	*Class*	*Family*	*Genera*	*Local Name*	*Site–I*	*Site–II*	*Site III*	*Site–IV*
1.	Pteridophyta	Salvinaceae	Salvinia	Water fern	+	+	+	–
2.	Pteridophyta	Marsileaseae	Marselia	Water fern	–	+	+	–
3.	Dicotyledons	Nymphaeceae	Nelumbo	Lotus	–		+	–
4.	Dicotyledons	Trapaceae	Trapa	Singhara	–	+	+	–
5.	Dicotyledons	Convolvulaceae	Ipomoea	Morning glory	–	+	+	–
6.	Dicotyledons	Polygonaceae	Polygonum	Knot weed	–	+	+	–
7.	Dicotyledons	Hydrocharitaceae	Hydrilla	Knot weed	–	+	+	–
8.	Dicotyledons	Pontedericeae	Eicchornia	Water hyacinth	+	+	+	–
9.	Monocotyledon	Typhaceae	Typha (Cattail)	Elephant grass	–	+	+	–
10.	Monocotyledon	Araceae	Pistia	Duck weed	+	+	–	–
11.	Monocotyledon	Lemnaceae	Lemna	Lesser Duck Weed	+	–	–	+
12.	Monocotyledon	Potamogetonaceae	Potamogeton	Pond weed	+	+	+	+
13.	Monocotyledon	Cyperaceae	Cyperus	Bulrushes	–	+	+	–
14.	Monocotyledon	Araceae	Colocasia	Taro, Arum	–	+	+	–

Note: + : Present; –: Absent.

Table 5.III.C.i.1: Quantitative Analysis of Total Coliform Bacteria at Sampling Site–I (2007-09)

Months	*MPN/100 ML*	
	2007-2008	*2008-2009*
March	39.00	44.00
April	60.00	64.00
May	65.00	79.00
June	79.00	88.00
July	110.00	107.00
August	123.00	159.00
September	164.00	176.00
October	155.00	160.00
November	13.00	11.00
December	08.00	09.00
January	07.00	13.00
February	15.00	17.00
Average ±SD	60.83± 56.99	77.25± 62.07

SD ±: Standard Deviation.

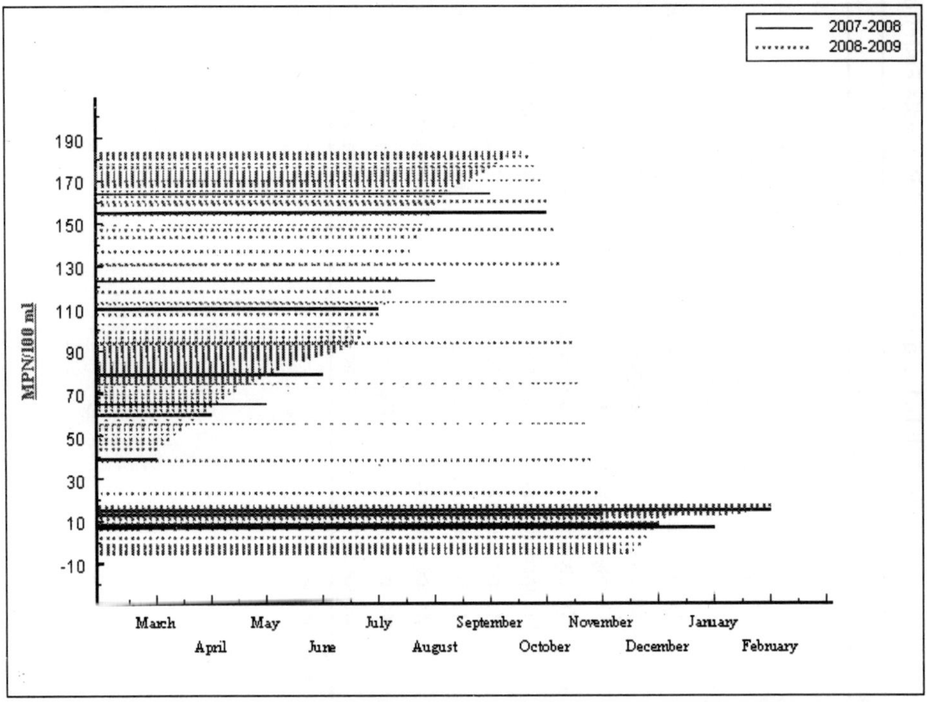

Figure 5.III.C.i.1: Graph Showing the Monthly Variation in MPN at Sampling Site–I (2007-2009)

Sampling Site–II

The MPN /100ml of the pond at this site was recorded minimum 7.00 MPN/ 100 ml in the month of January in the year 2007-2008 and maximum of 173.00 MPN/ 100 ml in the month of September in the year 2008-2009. The average value in the year 2007-2008 was 70.92 MPN/ 100 ml ± 58.70 and 71.83 MPN/ 100 ml ± 59.83 in the year 2008-2009 (Table 5.III.C.i.2 and Figure 5.III.C.i.2).

Table 5.III.C.i.2: Quantitative Analysis of Total Coliform Bacteria at Sampling Site–II (2007-2009)

Months	*MPN/100 ML*	
	2007-2008	*2008-2009*
March	43.00	40.00
April	54.00	57.00
May	57.00	64.00
June	84.00	75.00
July	136.00	111.00
August	170.00	155.00
September	159.00	173.00
October	99.00	141.00
November	20.00	17.00
December	13.00	10.00
January	07.00	09.00
February	09.00	10.00
Average ±SD	70.92± 58.70	71.33± 59.83

SD ±: Standard Deviation.

Sampling Site–III

The MPN/100ml of the fish pond at this spot was recorded minimum 8.00 MPN/ 100 ml in the month of February in the year 2007-2008 and maximum of 170.00 MPN/100 ml in the month of September in the year 2008-2009. The average value in the year 2007-2008 was 71.33 MPN/100 ml ± 57.85 and 73.67 MPN/ 100 ml ± 63.19 in the year 2007-2008 (Table 5.III.C.i.3 and Figure 5.III.C.i.3).

Sampling Site–IV

The MPN/100ml of the pond at this sampling spot was recorded minimum 6.00 MPN/100 ml in the month of December in the year 2008-2009 and maximum of 187.00 MPN/100 ml in the month of September in the year 2008-2009. The average value in the year 2007-2008 was 79.42 MPN/100 ml ± 63.40 and 76.58 MPN/ 100 ml ± 61.03 in the year 2008-2009 (Table 5.III.C.i.4 and Figure 5.III.C.i.4).

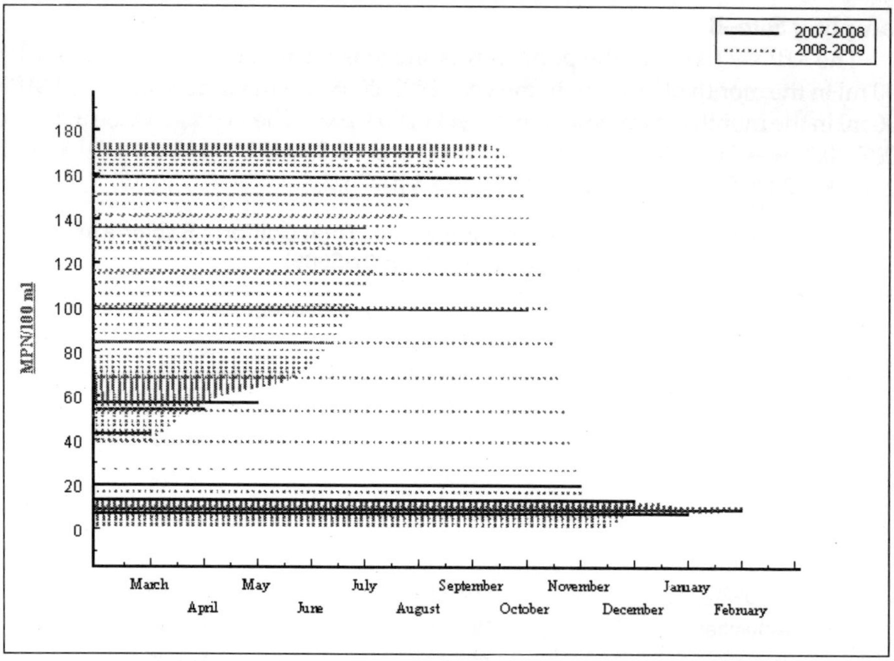

Figure 5.III.C.i.2 Graph Showing the Monthly Variation in MPN at Sampling Site–II (2007-2009)

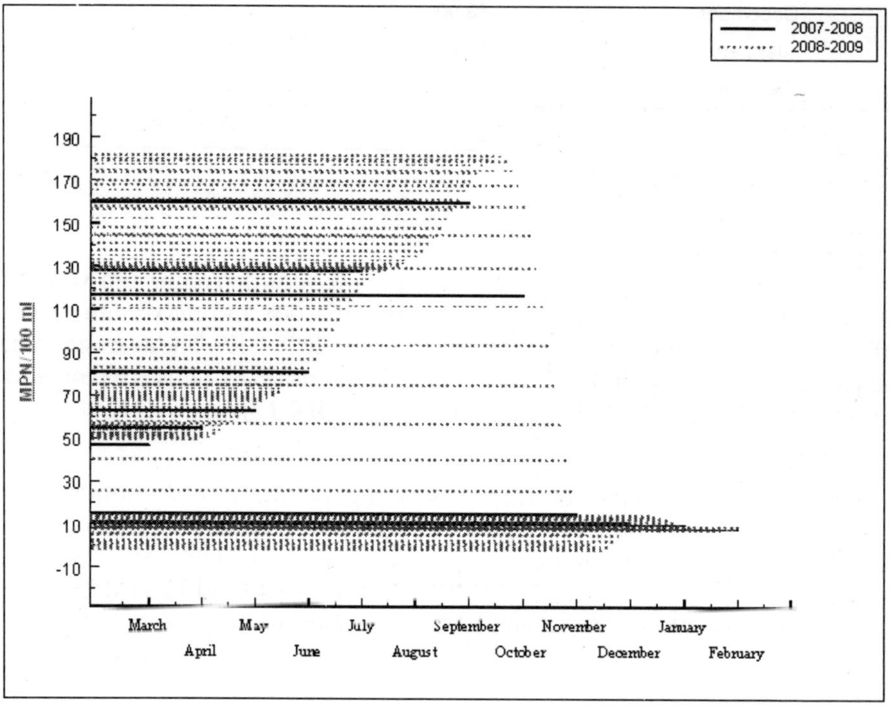

Figure 5.III.C.i.3: Graph Showing the Monthly Variation in MPN at Sampling Site–III (2007-2009)

Table 5.III.C.i.3: Quantitative Analysis of Total Coliform Bacteria at Sampling Site–III (2007-2009)

Months	*MPN/100 ML*	
	2007-2008	*2008-2009*
March	47.00	57.00
April	55.00	50.00
May	63.00	66.00
June	81.00	83.00
July	128.00	122.00
August	161.00	135.00
September	160.00	170.00
October	117.00	158.00
November	15.00	14.00
December	11.00	11.00
January	10.00	09.00
February	08.00	09.00
Average ±SD	71.33± 57.85	73.67± 63.19

SD ±: Standard Deviation.

Table: 5.III.C.i.4: Quantitative Analysis of Total Coliform Bacteria at Sampling Site–IV (2007-2009)

Months	*MPN/100 ML*	
	2007-2008	*2008-2009*
March	50.00	58.00
April	69.00	65.00
May	81.00	77.00
June	103.00	82.00
July	119.00	91.00
August	143.00	143.00
September	184.00	187.00
October	163.00	163.00
November	11.00	11.00
December	09.00	06.00
January	08.00	14.00
February	13.00	22.00
Average ±SD	79.42 ± 63.40	76.58± 61.03

SD ±: Standard Deviation.

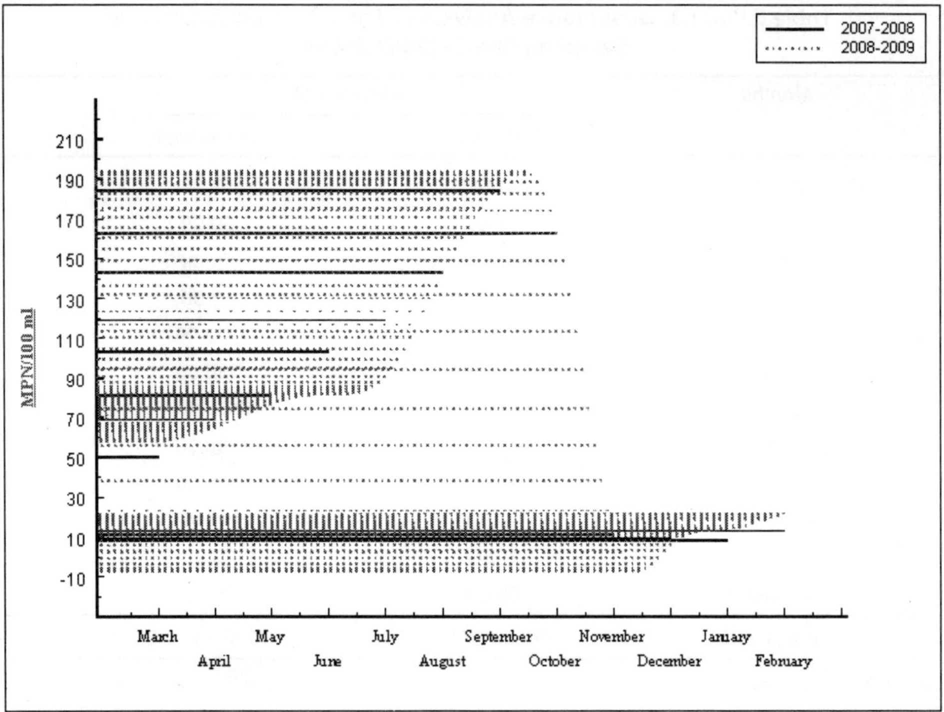

Figure 5.III.C.i.4: Graph Showing the Monthly Variation in MPN at Sampling Site–IV (2007-2009)

(ii) SPC (Standard Plate Count)

Sampling Site–I

The SPC/ml (x10^3) of Sampling site-I (Figure 5.1) was recorded minimum 6.00 x 10^3 SPC/ml in December month of the year 2007-2008 and maximum of 170.00 x 10^3 SPC/ml in the month of October in the year 2007-2008. The average value in the year 2007-2008 was 74.17 x 10^3 SPC/ml ± 61.93 and was recorded to be 70.83 x 10^3 SPC/ml ± 55.92 in the year 2006-2005. (Table 5.III.C.ii.1 and Figure 5.III.C.ii.1)

Sampling Site–II

The SPC/ml (x10^3) of Sampling site-II (Figure 5.2) was recorded minimum 8.00 x 10^3 SPC/ml in January month of the year 2007-2008 and maximum of 173.00 x 10^3 SPC/ml in the month of September in the year 2008-2009. The average value in the year 2007-2008 was 74.17 x 10^3 SPC/ml ± 61.83 and was recorded to be 75.75 x 10^3 SPC/ml ± 62.02 in the year 2008-2009. (Table 5.III.C.ii.2 and Figure 5.III.C.ii.2)

Sampling Site–III

The SPC/ml (x10^3) of sampling site-III (Figure 5.3) was recorded minimum 6.00 x 10^3 SPC/ml in February month of the year 2007-2008 and maximum of 170.00 x 10^3 SPC/ml in the month of September in the year 2007-2008. The average value in the year 2007-2008 was 75.08 x 10^3 SPC/ml ± 63.19 and was recorded to be 73.08 x 10^3 SPC/ml ± 59.41 in the year 2008-2009. (Table 5.III.C.ii.3 and Figure 5.III.C.ii.3)

Table 5.III.C.ii.1: Quantitative Analysis of Standard Plate Count at Sampling Site–I (2007-2009)

Months	*SPC × 10³*	
	2007-2008	*2008-2009*
March	40.00	52.00
April	63.00	59.00
May	59.00	64.00
June	76.00	72.00
July	104.00	110.00
August	158.00	137.00
September	167.00	163.00
October	170.00	145.00
November	19.00	09.00
December	06.00	06.00
January	12.00	14.00
February	16.00	19.00
Average ±SD	74.17± 61.93	70.83± 55.92

SD ±: Standard Deviation.

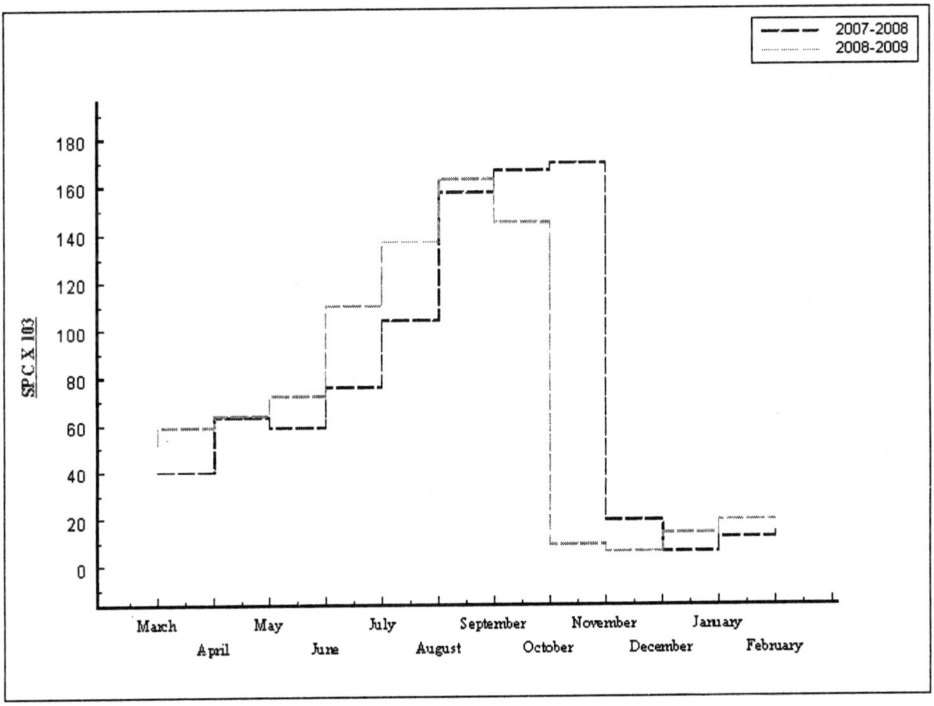

Figure 5.III.C.ii.1: Graph Showing the Monthly Variation in SPC at Sampling Site–I (2007-2009)

Table 5.III.C.ii.2: Quantitative Analysis of Standard Plate Count at Sampling Site–II (2007-2009)

Months	*SPC × 10³*	
	2007-2008	*2008-2009*
March	45.00	54.00
April	52.00	58.00
May	60.00	72.00
June	81.00	81.00
July	125.00	108.00
August	171.00	169.00
September	164.00	173.00
October	144.00	150.00
November	20.00	13.00
December	11.00	09.00
January	08.00	10.00
February	09.00	12.00
Average ±SD	74.17± 61.83	75.75± 62.02

SD ±: Standard Deviation.

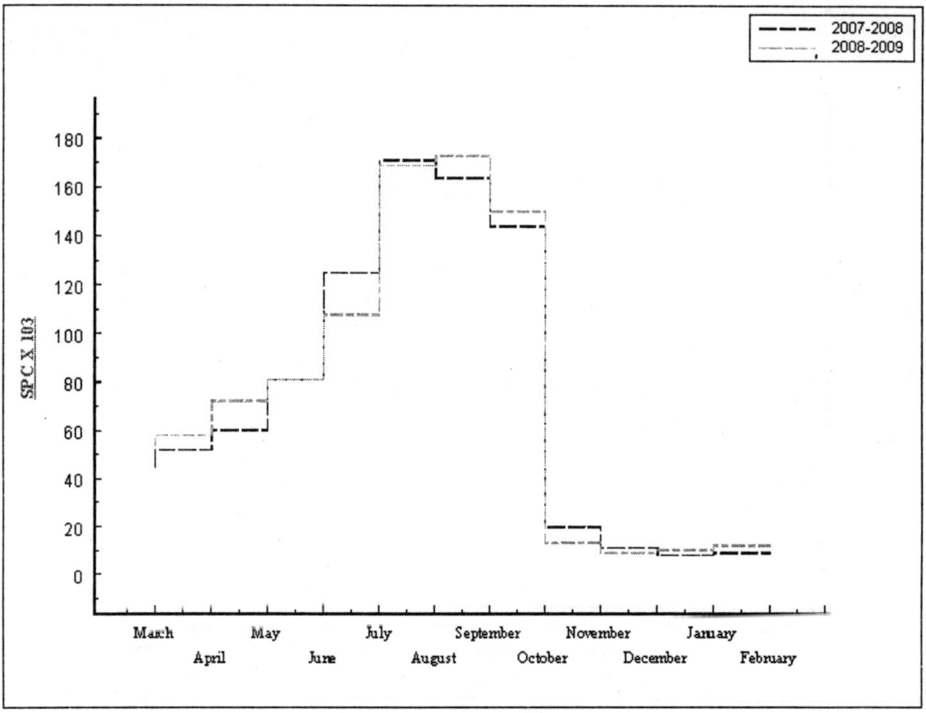

Figure 5.III.C.ii.2: Graph Showing the Monthly Variation in SPC at Sampling Site–II (2007-2009)

Table 5.III.C.ii.3: Quantitative Analysis of Standard Plate Count at Sampling Site–III (2007-09)

Months	*SPC x 10^3*	
	2007-2008	*2008-2009*
March	41.00	55.00
April	57.00	52.00
May	64.00	69.00
June	84.00	84.00
July	134.00	122.00
August	168.00	164.00
September	170.00	151.00
October	141.00	143.00
November	18.00	10.00
December	10.00	10.00
January	08.00	09.00
February	06.00	08.00
Average ±Sd	75.08± 63.19	73.08± 59.41

SD ±: Standard Deviation.

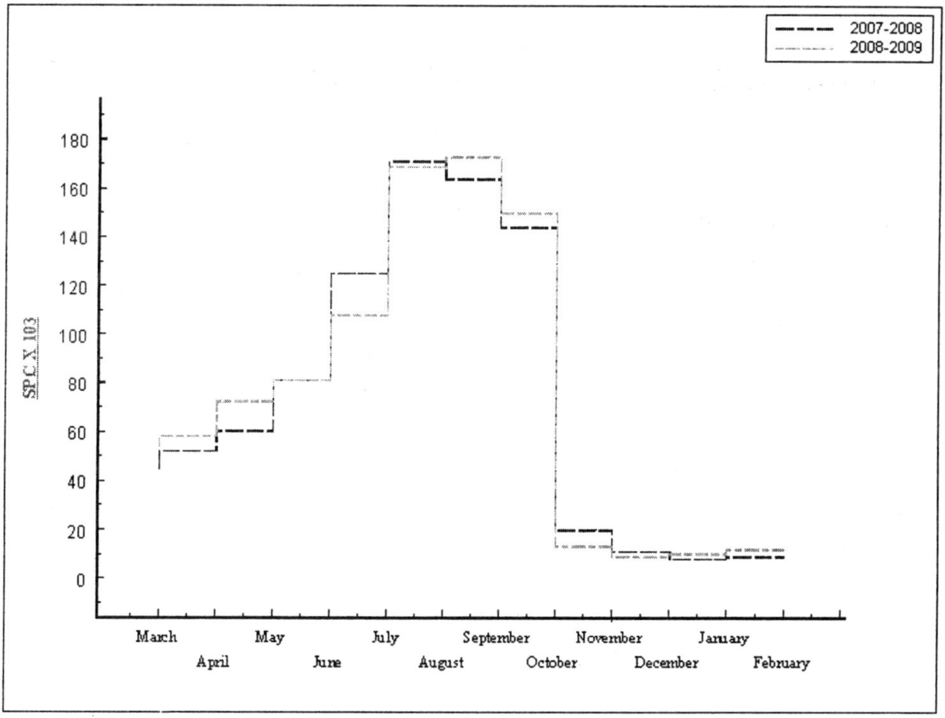

Figure 5.III.C.ii.3: Graph Showing the Monthly Variation in SPC at Sampling Site–III (2007-2009)

Table 5.III.C.ii.4: Quantitative Analysis of Standard Plate Count at Sampling Site–IV (2007-2009)

Months	*SPC x 10³*	
	2007-2008	*2008-2009*
March	43.00	43.00
April	58.00	66.00
May	60.00	73.00
June	77.00	88.00
July	110.00	137.00
August	154.00	148.00
September	193.00	181.00
October	180.00	151.00
November	20.00	134.00
December	13.00	19.00
January	17.00	11.00
February	22.00	25.00
Average ±Sd	78.92± 65.24	89.67± 58.85

SD ±: Standard Deviation.

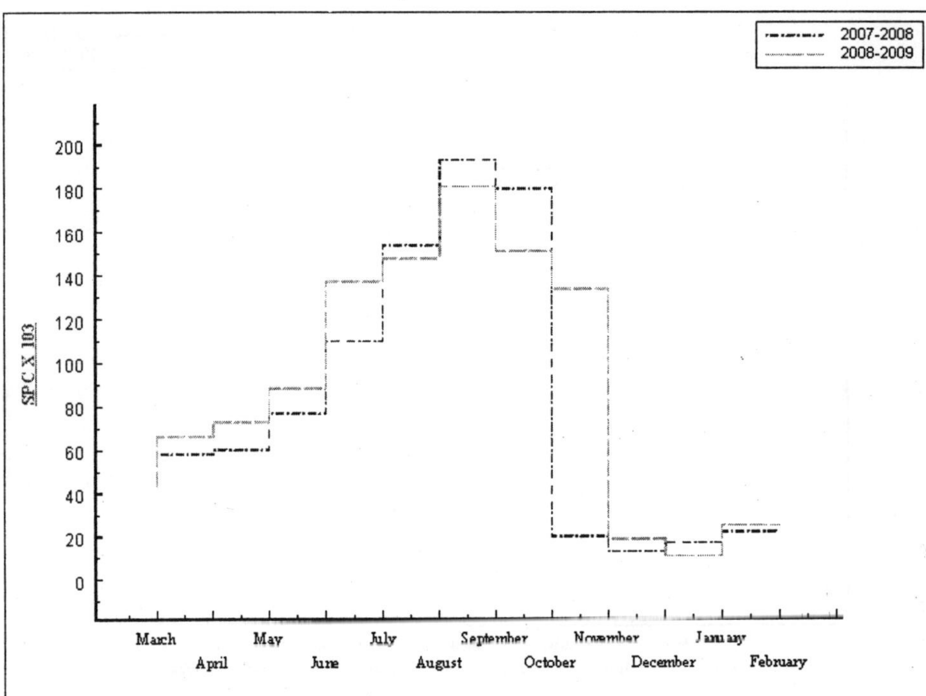

Figure 5.III.C.ii.4: Graph Showing the Monthly Variation in SPC at Sampling Site–IV (2007-2009)

Sampling Site–IV

The SPC/ml (x10^3) of sampling site-IV (Figure 5.4), was recorded minimum 11.00 x 10^3 SPC/ml in January month of the year 2008-2009 and maximum of 193 x 10^3 SPC/ml in the month of September in the year 2007-2008. The average value in the year 2007-2008 was 78.92 x 10^3 SPC/ml ± 65.24 and was recorded to be 89.67 x 10^3 SPC/ml ± 58.85 in the year 2006-20057. (Table 5.III.C.ii.4 and Figure 5.III.C.ii.4)

(iii) Aquatic Fungi

Although soil is by far the most common habitat of the fungi as a whole, many are aquatic. These fungi are known collectively as *water moulds* or *aquatic phycomycetes*. They occur on the surface of decaying plant or animal materials in aquatic bodies. Site wise details of the observations of fungi are mentioned in Table 5.III.C.iii.1 to 5.III.C.iii.4 and Figures 5.III.C.iii.1 to 5.III.C.iii.12.

Sampling Site-I

The result of aquatic fungi for this site is given in Table 5.III.C.iii.1 and in Figure 5.III.C.iii.1 to 5.III.C.iii.3.

During the course of study at study site – I (Figure 5.5), the Total aquatic fungi was noted lowest of 06 ml^{-1}x 10^3 in the month of July of 2007-2008 and highest of 48 ml^{-1}x 10^3 in October month of 2008-2009 with the average value for fungi was 24.33 ml^{-1}x 10^3 ± 12.37 in 2007-2008 and 26.17 ml^{-1}x 10^3 ±11.81 in 2008-2009 at sampling site-I.

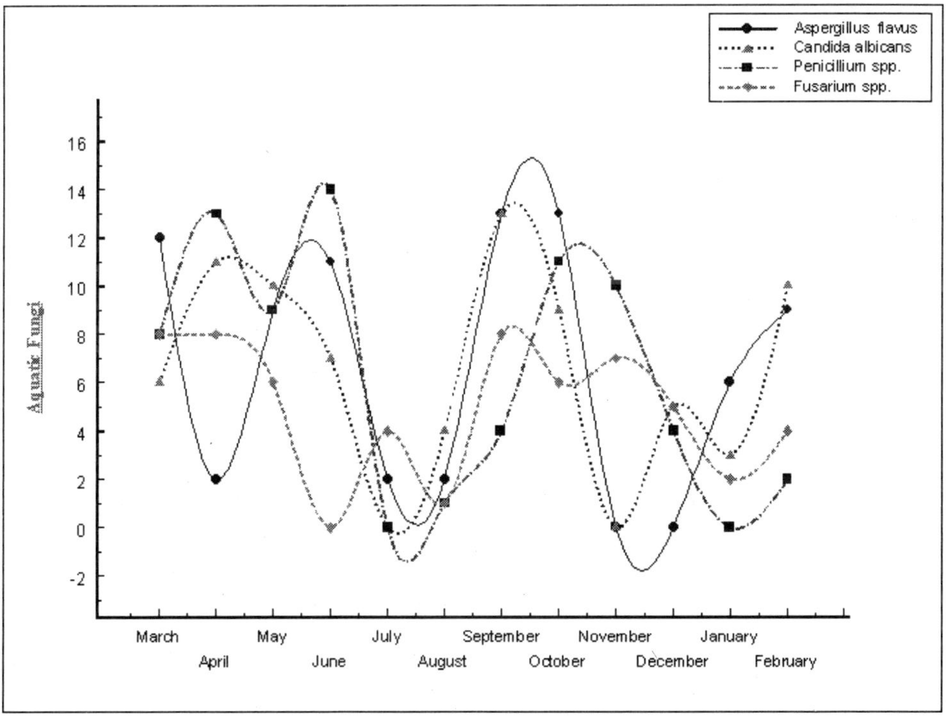

Figure 5.III.C.iii.1: Graph Showing the Monthly Variation in Aquatic Fungi at Sampling Site–I (2007-2008)

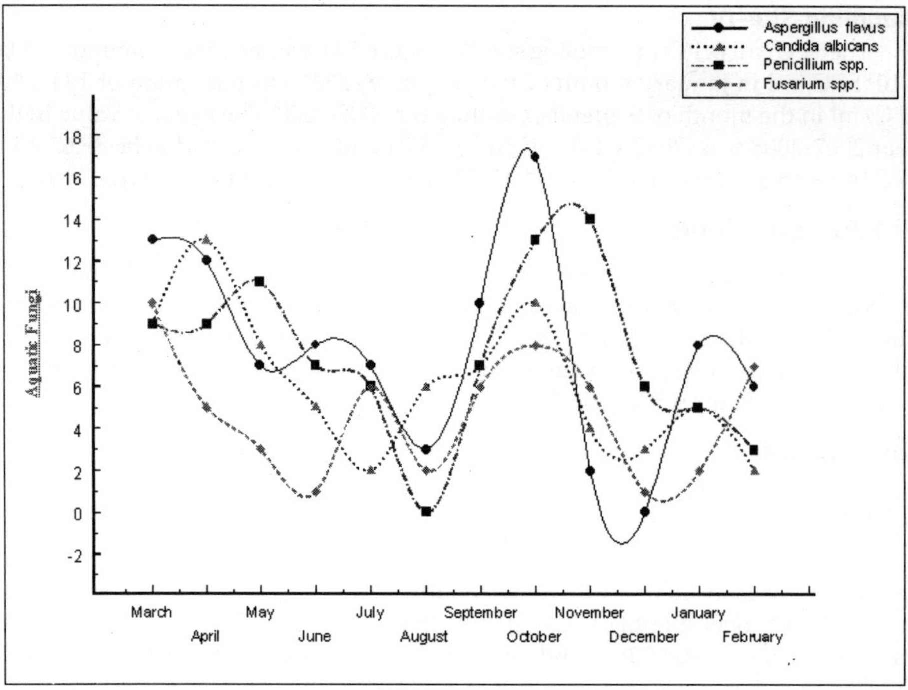

Figure 5.III.C.iii.2: Graph Showing the Monthly Variation in Aquatic Fungi at Sampling Site–I (2008-2009)

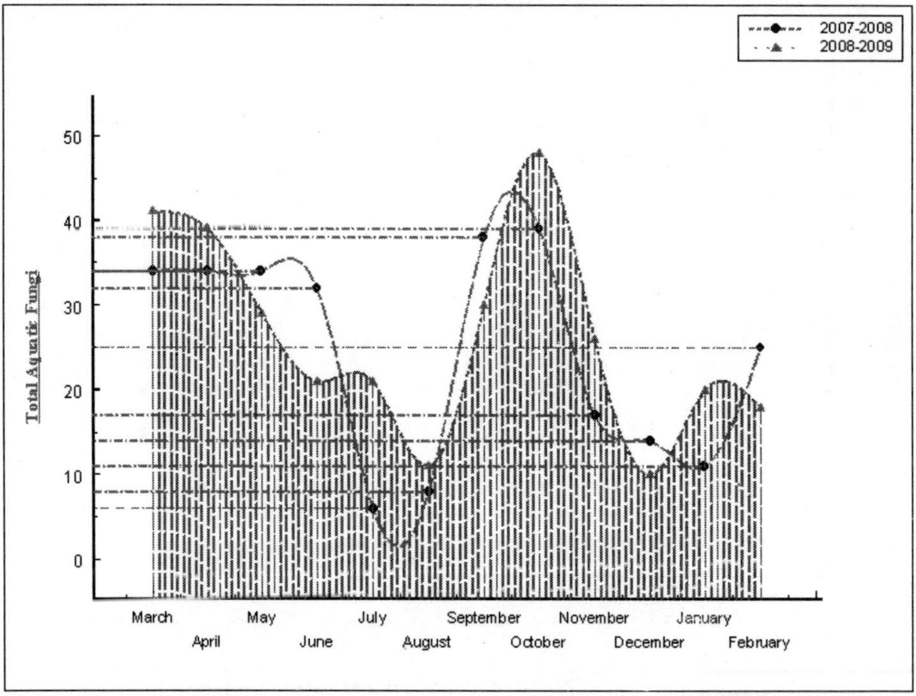

Figure 5.III.C.iii.3: Graph Showing the Monthly Variation in Total Aquatic Fungi at Sampling Site–I (2007-2009)

Table 5.III.C.iii.1: Qualitative and Quantitative Estimation for Aquatic Fungi from Sampling Site–I (2007-2009)

Months	*Aspergillus flavus*	*Candida albicans*	*Penicillium spp.*	*Fusarium spp.*	*Total ($ml^{-1}x10^3$)*
			(2007-2008)		
March	12.00	06.00	08.00	08.00	34.00
April	02.00	11.00	13.00	08.00	34.00
May	09.00	10.00	09.00	06.00	34.00
June	11.00	07.00	14.00	00.00	32.00
July	02.00	00.00	00.00	04.00	06.00
August	02.00	04.00	01.00	01.00	08.00
September	13.00	13.00	04.00	08.00	38.00
October	13.00	09.00	11.00	06.00	39.00
November	00.00	00.00	10.00	07.00	17.00
December	00.00	05.00	04.00	05.00	14.00
January	06.00	03.00	00.00	02.00	11.00
February	09.00	10.00	02.00	04.00	25.00
Average ±SD	6.58±5.16	6.50±4.25	6.33±5.11	4.92±2.78	24.33±12.37
			(2008-2009)		
March	13.00	09.00	09.00	10.00	41.00
April	12.00	13.00	09.00	05.00	39.00
May	07.00	08.00	11.00	03.00	29.00
June	08.00	05.00	07.00	01.00	21.00
July	07.00	02.00	06.00	06.00	21.00
August	03.00	06.00	00.00	02.00	11.00
September	10.00	07.00	07.00	06.00	30.00
October	17.00	10.00	13.00	08.00	48.00
November	02.00	04.00	14.00	06.00	26.00
December	00.00	03.00	06.00	01.00	10.00
January	08.00	05.00	05.00	02.00	20.00
February	06.00	02.00	03.00	07.00	18.00
Average ±SD	7.75±4.83	6.17±3.38	7.50±4.01	4.75±2.93	26.17±11.81

SD ±: Standard Deviation.

For aquatic fungi *Aspergillus flavus* was dominated followed by *Candida albicans, Penicillium* spp. and *Fusarium* spp. in both the years of study. At this sampling site *Aspergillus flavus* was observed maximum in the month of October of 2008-2009 (17 $ml^{-1}x 10^3$) and observed nil in November, December months of 2007-2008 and December month of 2008-2009, with average count of colony in 2007-2008 was 6.58 $ml^{-1}x 10^3$ ± 5.16 and 7.75 $ml^{-1}x 10^3$ ± 4.83 in 2008-2009. *Candida albicans* were found to be highest in the month of April 2008-2009 (13 ml^{-1}x 10^{3}) and lowest in 2007-2008 in the months of July and November (00 ml^{-1}x 10^{3}). The average count of *Candida albicans* in 2007-2008 was 6.50 ml^{-1}x 10^{3} ± 4.25 and 6.17 ml^{-1}x 10^{3} ± 3.38 in 2008-2009. *Penicillium* spp. showed optimum peak during the month of June in 2007-2008 (16 ml^{-1}x 10^{3}) and least peak during the month of July of 2007-2008 and August of 2008-2009 (00 ml^{-1}x 10^{3}). In 2007-2008 the average value of *Penicillium* spp. was 6.33 ml^{-1}x 10^{3} ± 5.11 and in 2008-2009 it was 7.50 ml^{-1}x 10^{3} ± 4.01. *Fusarium* spp. were found to be highest in the month of March 2008-2009 (10 ml^{-1}x 10^{3}) and recorded nil in 2007-2008 in the month of June. The average count of this species in 2007-2008 was 4.92 ml^{-1}x 10^{3} ± 2.78 and 4.75 ml^{-1}x 10^{3} ± 2.93 in 2008-2009.

Sampling Site–II

The result of aquatic fungi for this site is given in Table 5.III.C.iii.2 and in Figure 5.III.C.iii.4 to 5.III.C.iii.6.

At sampling site–IV (Figure 5.2) the Total aquatic fungi was noted minimum of 14 ml^{-1}x 10^{3} in the month of August of 2007-2008 and maximum of 45 ml^{-1}x 10^{3} in April month of the same year. The average value for fungi was 29.42 ml^{-1}x 10^{3} ± 12.37 in 2007-2008 and 24.67 ml^{-1}x 10^{3} ± 6.75 in 2008-2009.

For *Aspergillus flavus* the value was lowest in the month of November of 2008-2009 (01 ml^{-1}x 10^{3}) and observed highest in June month of 2007-2008 (15 ml^{-1}x 10^{3}) with average value of 8.08 ml^{-1}x 10^{3} ± 4.03 in 2007-2008 was and 7.50 ml^{-1}x 10^{3} ± 4.40 in 2008-2009. *Candida albicans* were found to be highest in the month of April 2008-

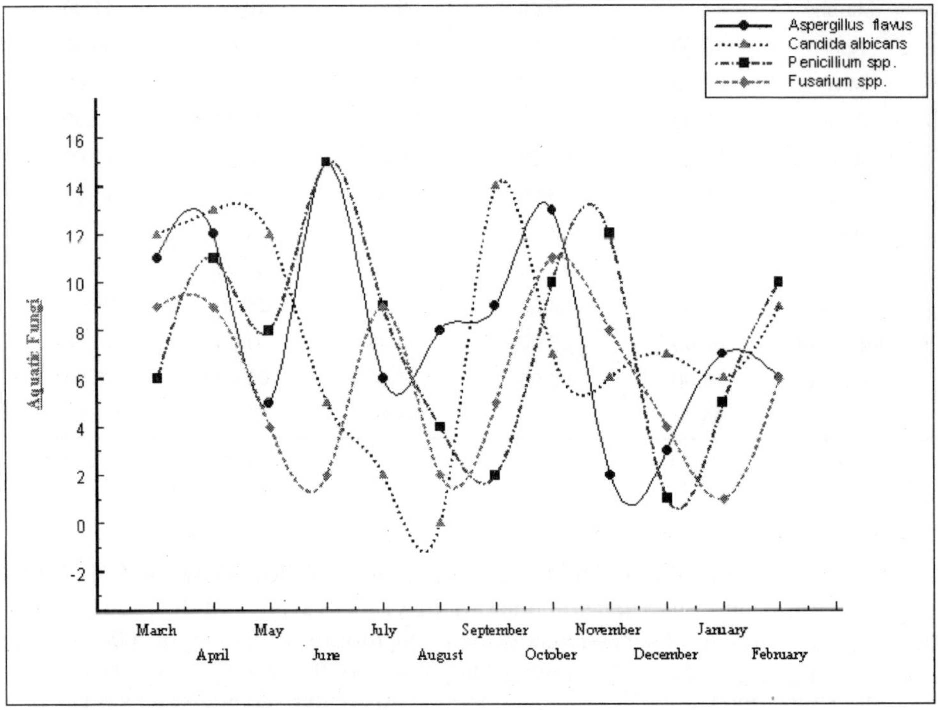

Figure 5.III.C.iii.4: Graph Showing the Monthly Variation in Aquatic Fungi at Sampling Site–II (2007-2008)

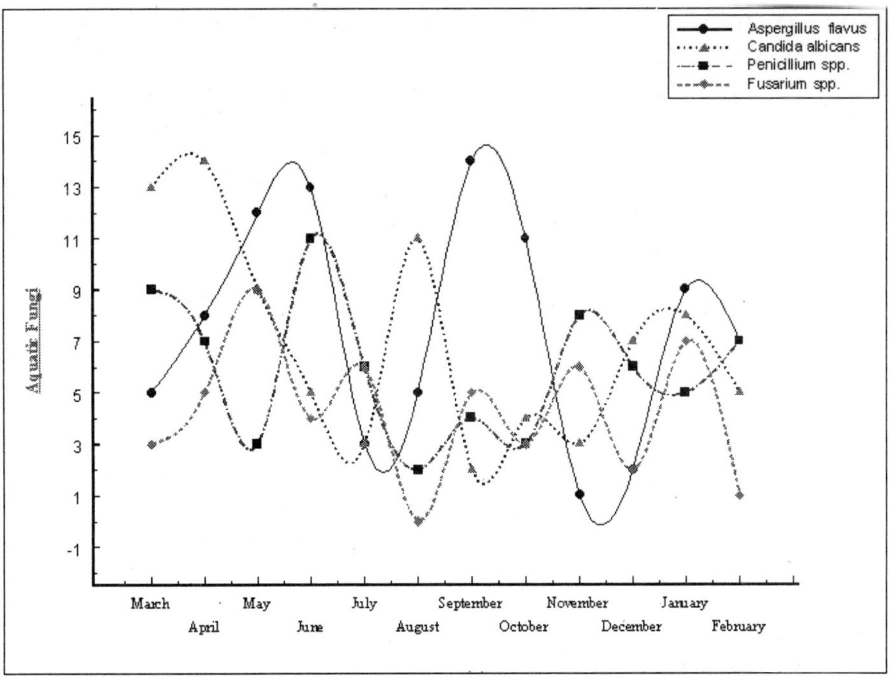

Figure 5.III.C.iii.5: Graph Showing the Monthly Variation in Aquatic Fungi at Sampling Site–II (2008-2009)

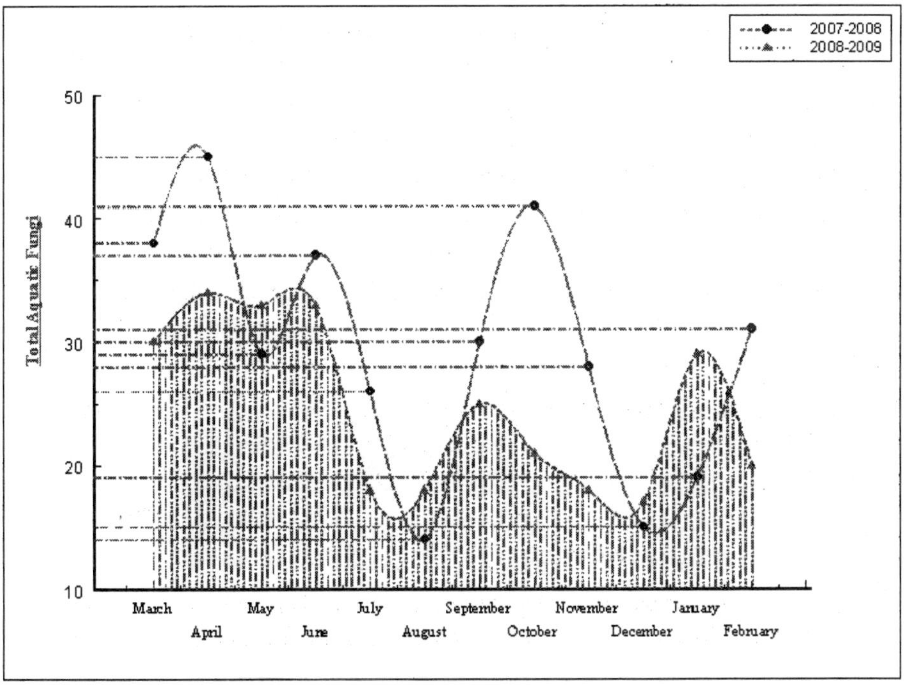

Figure 5.III.C.iii.6: Graph Showing the Monthly Variation in Total Aquatic Fungi at Sampling Site–II (2007-2009)

2009 (14 ml^{-1}x 10^3) and lowest in 2007-2008 in the month of August (00 ml^{-1}x 10^3). The average colony count for *Candida albicans* in 2007-2008 was 7.75 ml^{-1}x 10^3 ± 4.39 and 7.00 ml^{-1}x 10^3 ± 4.05 in 2008-2009. *Penicillium* spp. showed optimum peak during the month of June in 2007-2008 (15 ml^{-1}x 10^3) and least peak during the month of December of 2007-2008 (01 ml^{-1}x 10^3). In 2007-2008 the average value of *Penicillium* spp. was 7.75 ml^{-1}x 10^3 ± 4.22 and in 2008-2009 it was 5.92 ml^{-1}x 10^3 ± 2.68. *Fusarium* spp. were found to be highest in the month of October 2007-2008 (11 ml^{-1}x 10^3) and recorded lowest in 2008-2009 in the month of August (00 ml^{-1}x 10^3). The average count of this species in 2007-2008 was 5.83 ml^{-1}x 10^3 ± 3.33 and 4.25 ml^{-1}x 10^3 ± 2.60 in 2008-2009.

Table 5.III.C.iii.2: Qualitative and Quantitative Estimation for Aquatic Fungi from Sampling Site-II (2007-2009)

Months	*Aspergillus flavus*	*Candida albicans*	*Penicillium spp.*	*Fusarium spp.*	*Total (ml^{-1}x10^3)*
			(2007-2008)		
March	11.00	12.00	06.00	09.00	38.00
April	12.00	13.00	11.00	09.00	45.00
May	05.00	12.00	08.00	04.00	29.00
June	15.00	05.00	15.00	02.00	37.00
July	06.00	02.00	09.00	09.00	26.00
August	08.00	00.00	04.00	02.00	14.00
September	09.00	14.00	02.00	05.00	30.00
October	13.00	07.00	10.00	11.00	41.00
November	02.00	06.00	12.00	08.00	28.00
December	03.00	07.00	01.00	04.00	15.00
January	07.00	06.00	05.00	01.00	19.00
February	06.00	09.00	10.00	06.00	31.00
Average ±SD	8.08±4.03	7.75±4.39	7.75±4.22	5.83±3.33	29.42±12.37
			(2008-2009)		
March	05.00	13.00	09.00	03.00	30.00
April	08.00	14.00	07.00	05.00	34.00
May	12.00	09.00	03.00	09.00	33.00
June	13.00	05.00	11.00	04.00	33.00
July	03.00	03.00	06.00	06.00	18.00
August	05.00	11.00	02.00	00.00	18.00
September	14.00	02.00	04.00	05.00	25.00
October	11.00	04.00	03.00	03.00	21.00
November	01.00	03.00	08.00	06.00	18.00
December	02.00	07.00	06.00	02.00	17.00
January	09.00	08.00	05.00	07.00	29.00
February	07.00	05.00	07.00	01.00	20.00
Average ±SD	7.50±4.40	7.00±4.05	5.92±2.68	4.25±2.60	24.67±6.75

SD ±: Standard Deviation.

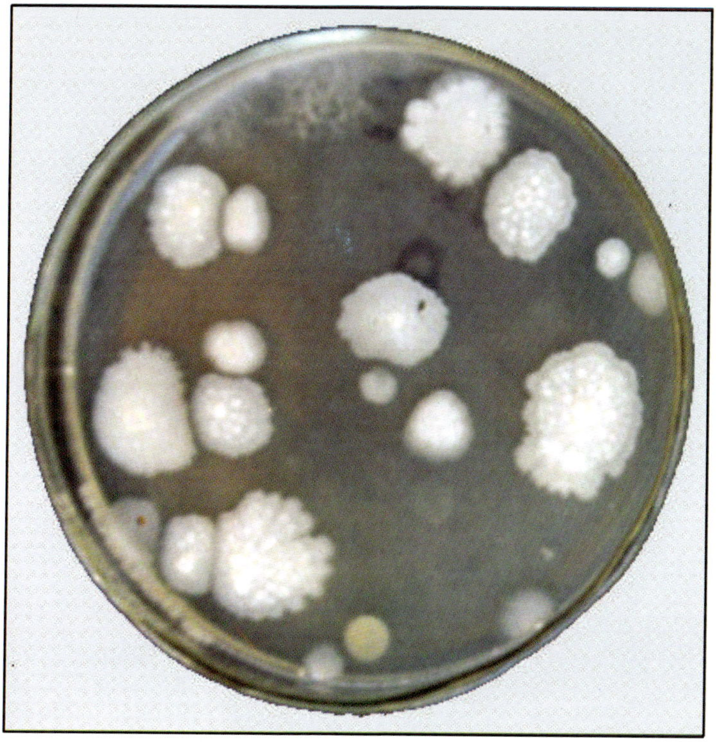

Figure 5.1: Showing SPC at Sampling Site–I

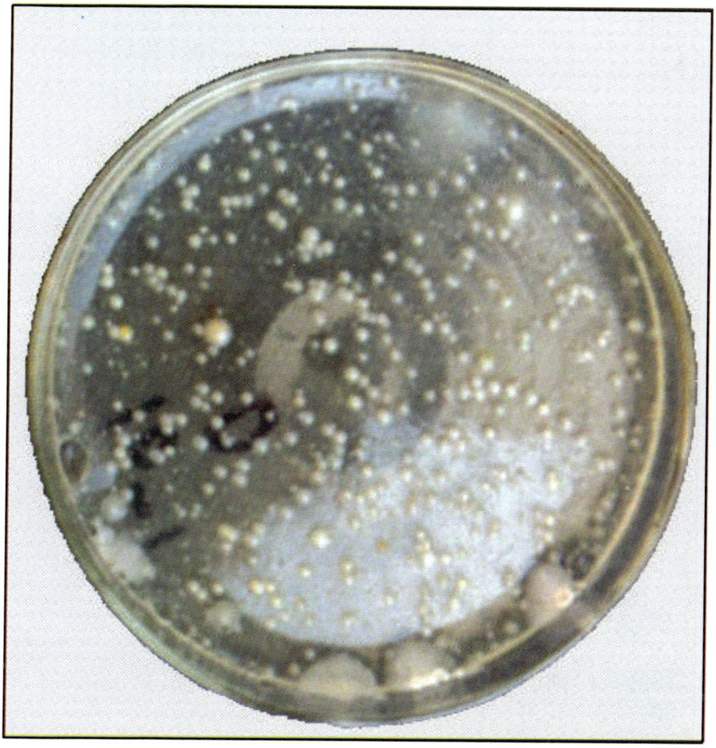

Figure 5.2: Showing SPC at Sampling Site–II

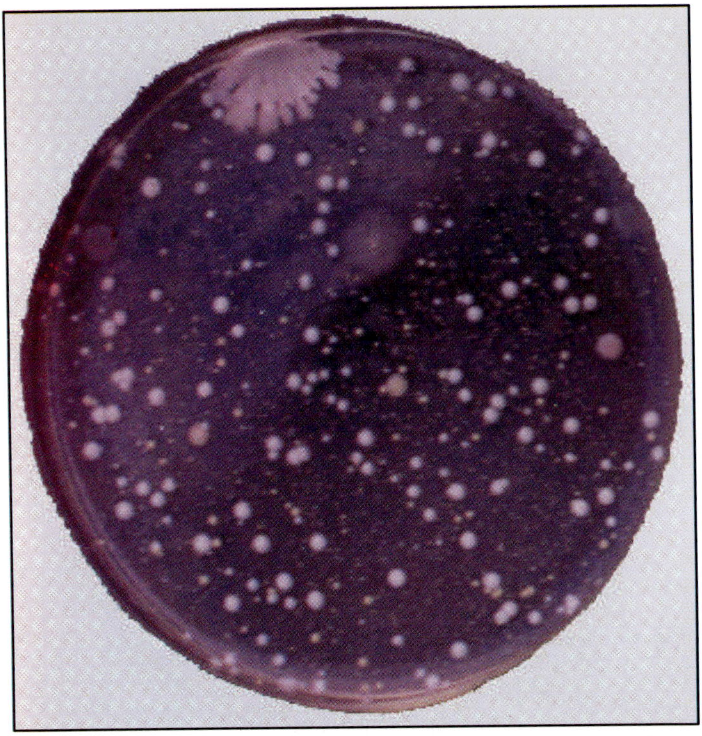

Figure 5.3: Showing SPC at Sampling Site–III

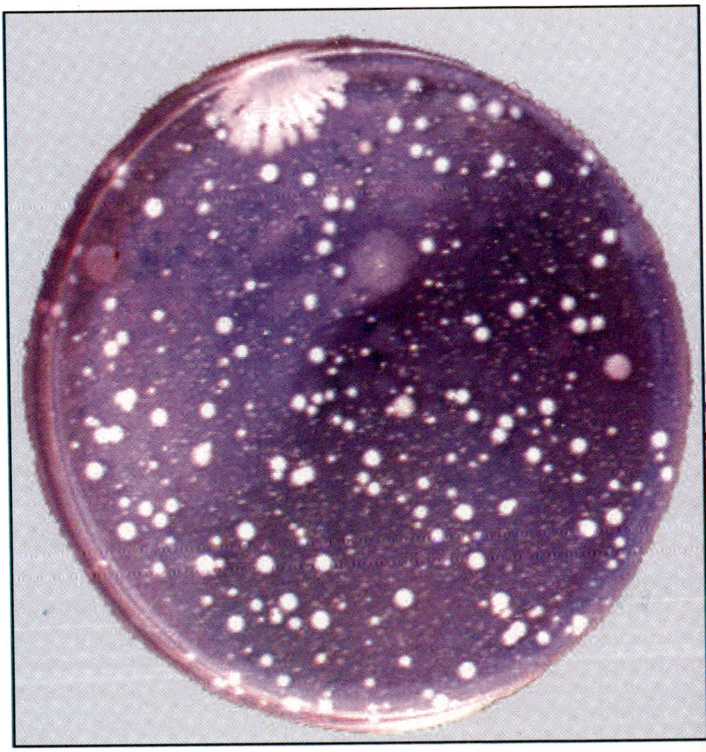

Figure 5.4: Showing SPC at Sampling Site–IV

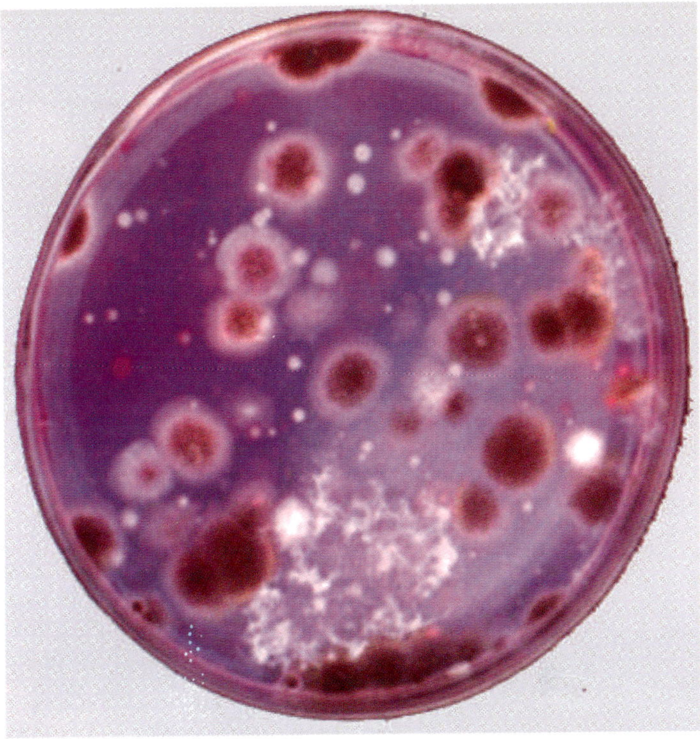

Figure 5.5: Showing Aquatic Fungi at Sampling Site–I

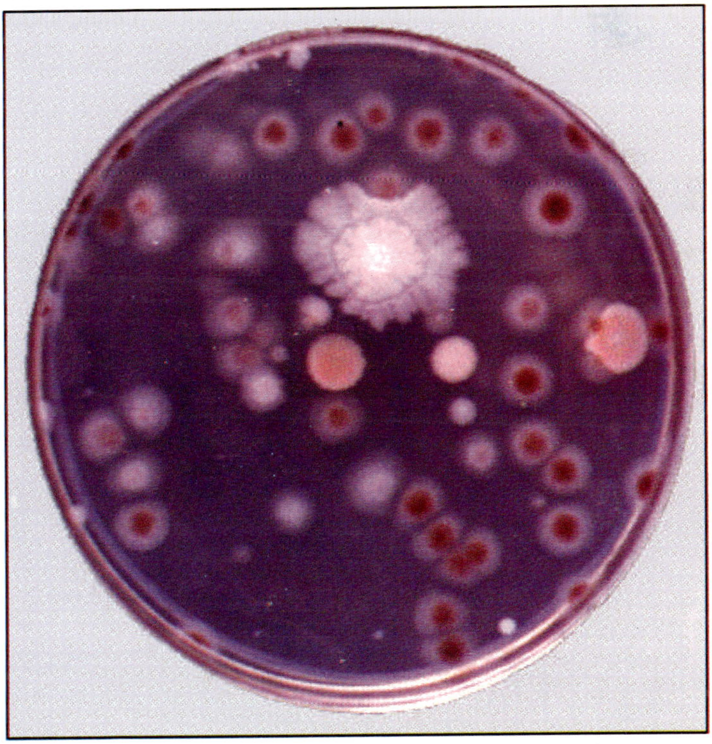

Figure 5.6: Showing Aquatic Fungi at Sampling Site–II

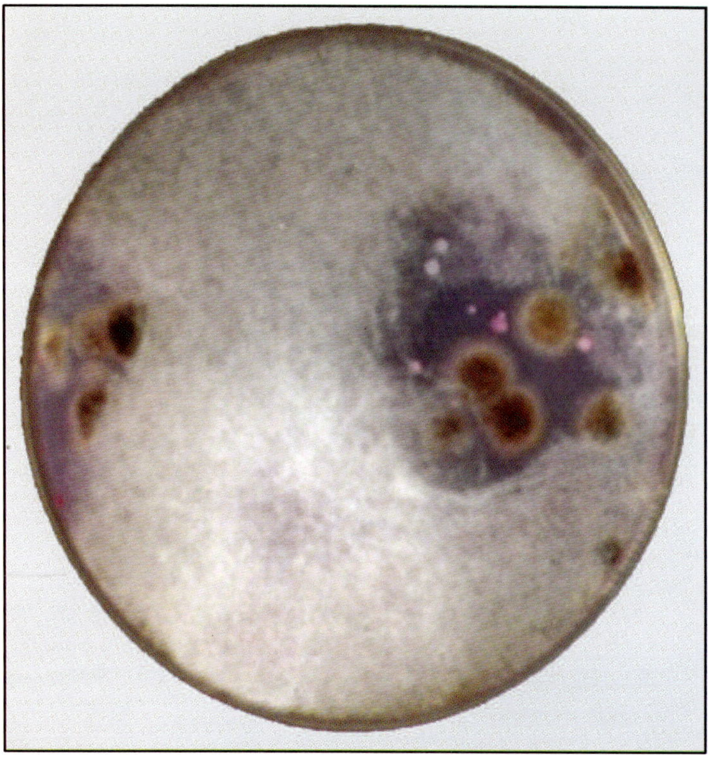

Figure 5.7: Showing Aquatic Fungi at Sampling Site–III

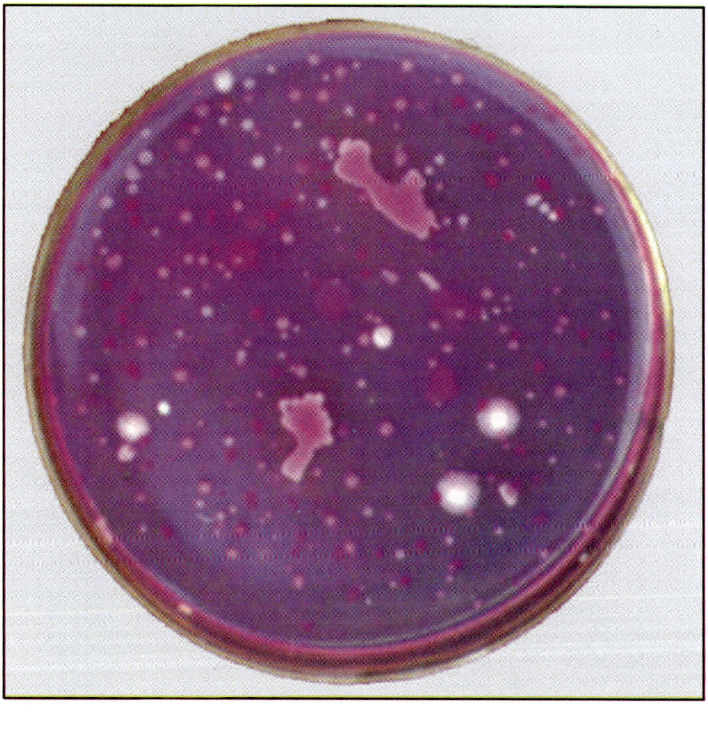

Figure 5.8: Showing Aquatic Fungi at Sampling Site–IV

Sampling Site–III

The result of aquatic fungi for this site is given in Table 5.III.C.iii.3 and in Figure 5.III.C.iii.7 to 5.III.C.iii.9.

In our study of Paniyala Fish pond at sampling site – III (Figure 5.7), the Total fungi was noted lowest of 09 $ml^{-1}x\ 10^3$ in the month of November of 2007-2008 and highest of 42 $ml^{-1}x\ 10^3$ in March month of 2008-2009 with the average value for fungi was 26.42 $ml^{-1}x\ 10^3 \pm 9.77$ in 2007-2008 and 26.92 $ml^{-1}x\ 10^3 \pm 7.87$ in 2008-2009.

Table 5.III.C.iii.3: Qualitative and Quantitative Estimation for Aquatic Fungi from Sampling Site–III (2007-2009)

Months	*Aspergillus flavus*	*Candida albicans*	*Penicillium spp.*	*Fusarium spp.*	*Total ($ml^{-1}x10^3$)*
			(2007-2008)		
March	09.00	10.00	11.00	08.00	38.00
April	04.00	12.00	10.00	02.00	28.00
May	03.00	12.00	07.00	05.00	27.00
June	10.00	09.00	01.00	03.00	23.00
July	09.00	03.00	04.00	02.00	18.00
August	11.00	05.00	11.00	05.00	32.00
September	15.00	14.00	06.00	06.00	41.00
October	13.00	08.00	10.00	04.00	35.00
November	01.00	01.00	04.00	03.00	0S9.00
December	02.00	09.00	06.00	01.00	18.00
January	03.00	07.00	04.00	02.00	16.00
February	14.00	11.00	03.00	04.00	32.00
Average ±Sd	7.83±5.01	8.42±3.87	6.42±3.40	3.75±2.01	26.42±9.77
			(2008-2009)		
March	10.00	14.00	07.00	08.00	39.00
April	10.00	09.00	12.00	10.00	41.00
May	06.00	10.00	10.00	08.00	34.00
June	12.00	03.00	08.00	03.00	26.00
July	04.00	04.00	07.00	07.00	22.00
August	07.00	02.00	05.00	05.00	19.00
September	06.00	11.00	03.00	06.00	26.00
October	11.00	02.00	11.00	09.00	33.00
November	03.00	04.00	08.00	04.00	19.00
December	05.00	09.00	03.00	06.00	23.00
January	08.00	04.00	06.00	03.00	21.00
February	04.00	03.00	09.00	04.00	20.00
Average ±Sd	7.17±3.01	6.25±4.09	7.42±2.88	6.08±2.35	26.92±7.87

SD ±: Standard Deviation.

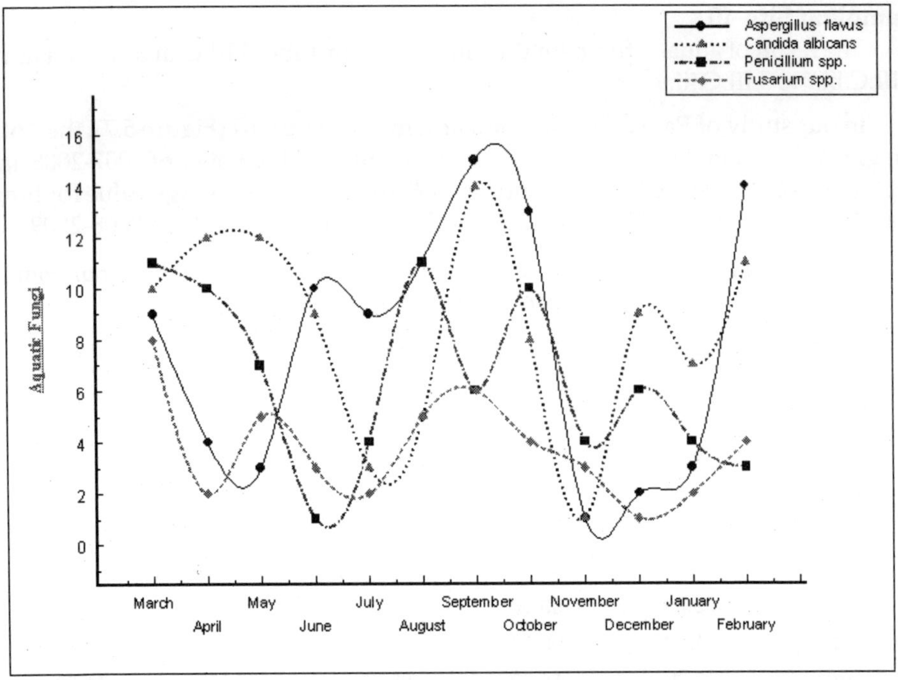

Figure 5.III.C.iii.7: Graph Showing the Monthly Variation in Aquatic Fungi at Sampling Site–III (2007-2008)

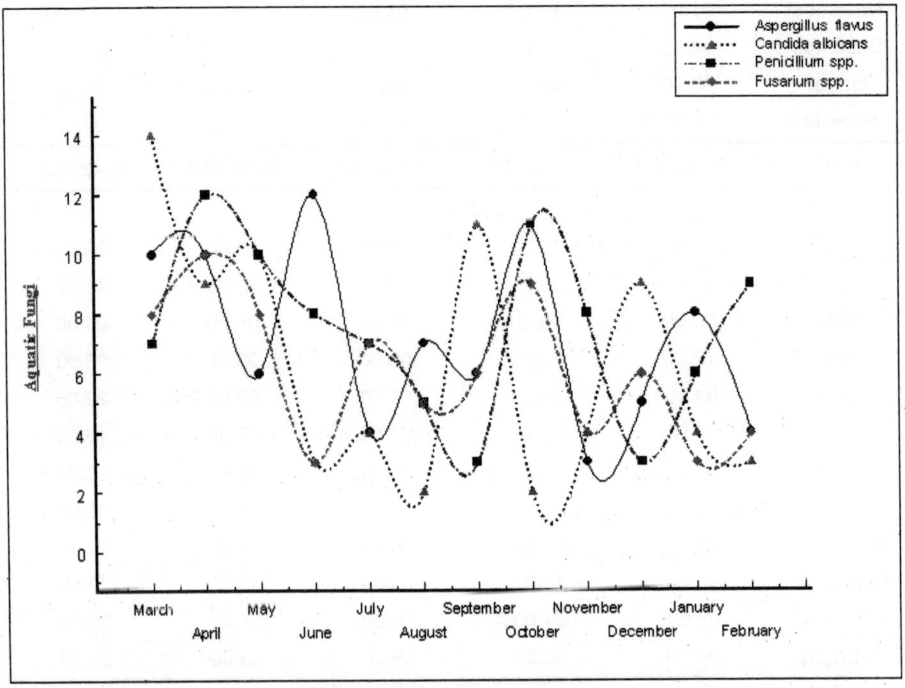

Figure 5.III.C.iii.8: Graph Showing the Monthly Variation in Aquatic Fungi at Sampling Site–III (2008-2009)

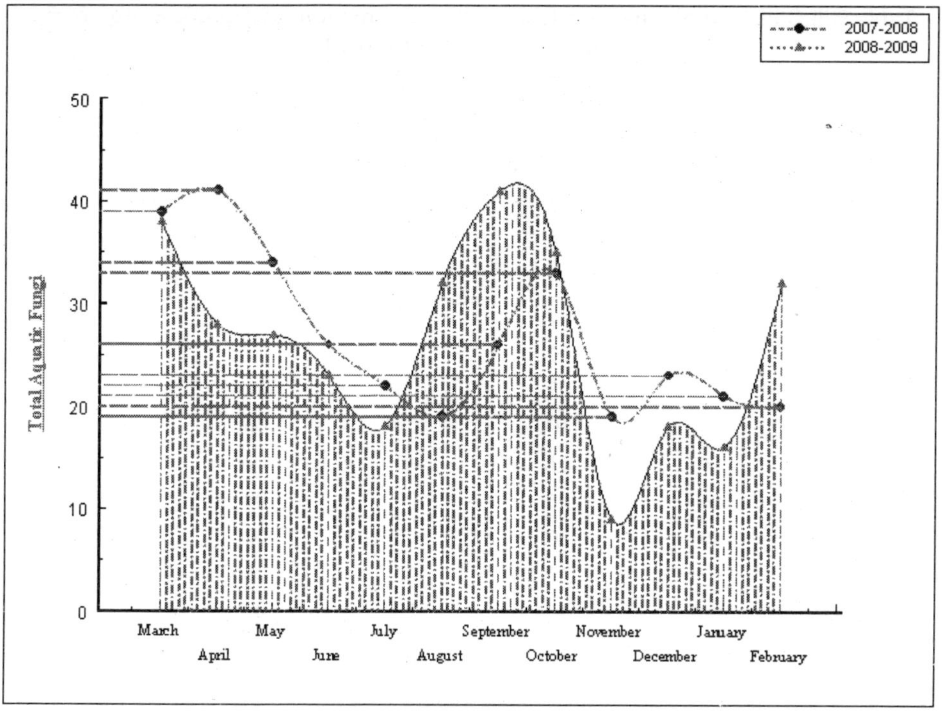

Figure 5.III.C.iii.9: Graph Showing the Monthly Variation in Total Aquatic Fungi at Sampling Site–III (2007-2009)

At this site, *Aspergillus flavus* was observed maximum in the month of September of 2007-2008 (15 ml^{-1}x 10^3) and observed 01 ml^{-1}x 10^3 in November month of 2007-2008, with average count of colony in 2007-2008 was 7.83 ml^{-1}x 10^3 ± 5.01 and 7.17 ml^{-1}x 10^3 ± 3.01 in 2008-2009. *Candida albicans* were found to be highest in the month of March 2008-2009 (14 ml^{-1}x 10^3) and lowest in 2007-2008 in the month of November (01 ml^{-1}x 10^3). The average count of *Candida albicans* in 2007-2008 was 8.42 ml^{-1}x 10^3 ± 3.87 and 6.25 ml^{-1}x 10^3 ± 4.09 in 2008-2009. *Penicillium* spp. showed optimum peak during the month of April in 2008-2009 (12 ml^{-1}x 10^3) and least peak during the month of June of 2007-2008 (01 ml^{-1}x 10^3). In 2007-2008 the average value of *Penicillium* spp. was 6.42 ml^{-1}x 10^3± 3.40 and in 2008-2009 it was 7.42 ml^{-1}x 10^3± 2.88. *Fusarium* spp. were found to be highest in the month of April 2008-2009 (10 ml^{-1}x 10^3) and recorded 01 ml^{-1}x 10^3 in 2007-2008 in the month of December. The average count of this species in 2007-2008 was 3.75 ml^{-1}x 10^3 ± 2.01 and 6.08 ml^{-1}x 10^3 ± 2.35 in 2008-2009.

Sampling Site–IV

The result of aquatic fungi for this site is given in Table 5.III.C.iii.4 and in Figure 5.III.C.iii.10 to 5.III.C.iii.12.

At sampling site-IV (Figure 5.8) the Total aquatic fungi was noted minimum of 15 ml^{-1}x 10^3 in the months of August and December of 2008-2009 and maximum of 45 ml^{-1}x 10^3 in April month of the same year. The average value for fungi was 28.75 ml^{-1}x 10^3 ± 7.41 in 2007-2008 and 27.83 ml^{-1}x 10^3 ± 9.68 in 2008-2009.

Table 5.III.C.iii.4: Qualitative and Quantitative Estimation for Aquatic Fungi from Sampling Site–IV (2007-2009)

Months	*Aspergillus flavus*	*Candida albicans*	*Penicillium spp.*	*Fusarium spp.*	*Total ($ml^{-1}x10^3$)*
		(2007-2008)			
March	13	03	10	07	33
April	13	08	10	09	40
May	11	12	08	05	36
June	14	06	15	02	37
July	09	04	02	05	20
August	06	09	00	02	17
September	10	10	02	05	27
October	14	01	10	07	32
November	06	05	09	09	29
December	08	06	05	02	21
January	07	07	03	05	22
February	05	15	07	04	31
Average ±Sd	9.67±3.31	7.17±3.93	6.75±4.41	5.17±2.48	28.75±7.41
		(2008-2009)			
March	12	10	07	07	36
April	11	10	13	11	45
May	06	13	05	03	27
June	14	06	13	01	34
July	05	01	07	08	21
August	09	03	02	01	15
September	04	10	09	06	29
October	12	06	12	10	40
November	05	08	08	09	30
December	01	09	02	03	15
January	06	05	06	02	19
February	03	07	08	05	23
Average ±Sd	7.33±4.14	7.33±3.37	7.67±3.73	5.50±3.53	27.83±9.68

SD ±: Standard Deviation.

For *Aspergillus flavus* the value was lowest in the month of December of 2008-2009 (01 $ml^{-1}x\ 10^3$) and observed highest in October month of 2007-2008 and June month in 2008-2009 (14 $ml^{-1}x\ 10^3$), with average value of 9.67 $ml^{-1}x\ 10^3 \pm 3.31$ in 2007-2008 was and 7.33 $ml^{-1}x\ 10^3 \pm 4.14$ in 2008-2009. *Candida albicans* were found to be highest in the month of February 2007-2008 (15 $ml^{-1}x\ 10^3$) and lowest in month of October in 2007-2008 and July 2008-2009 (01 $ml^{-1}x\ 10^3$). The average colony count for

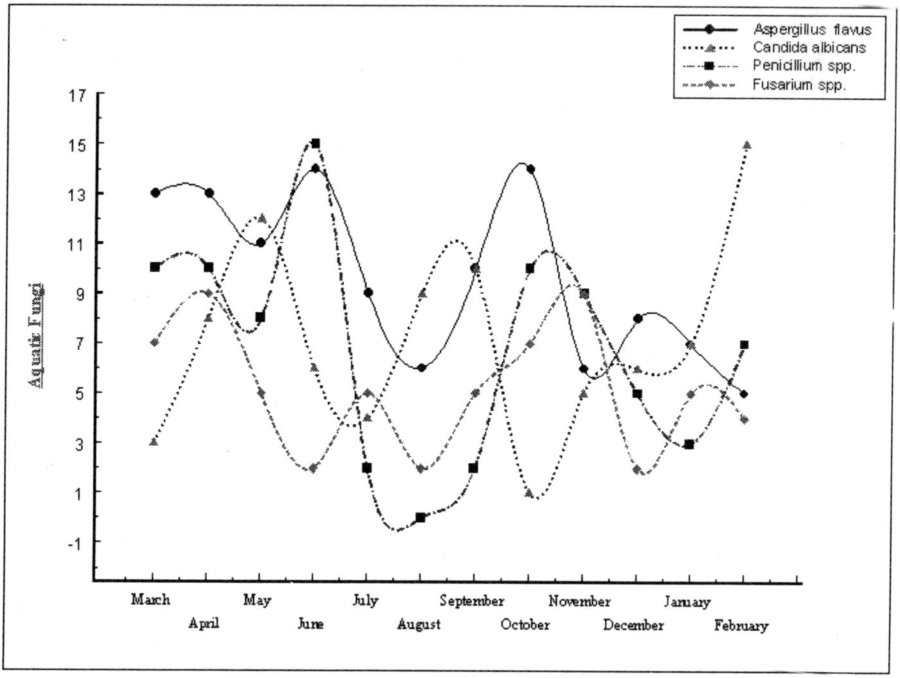

Figure 5.III.C.iii.10: Graph Showing the Monthly Variation in Aquatic Fungi at Sampling Site–IV (2007-2008)

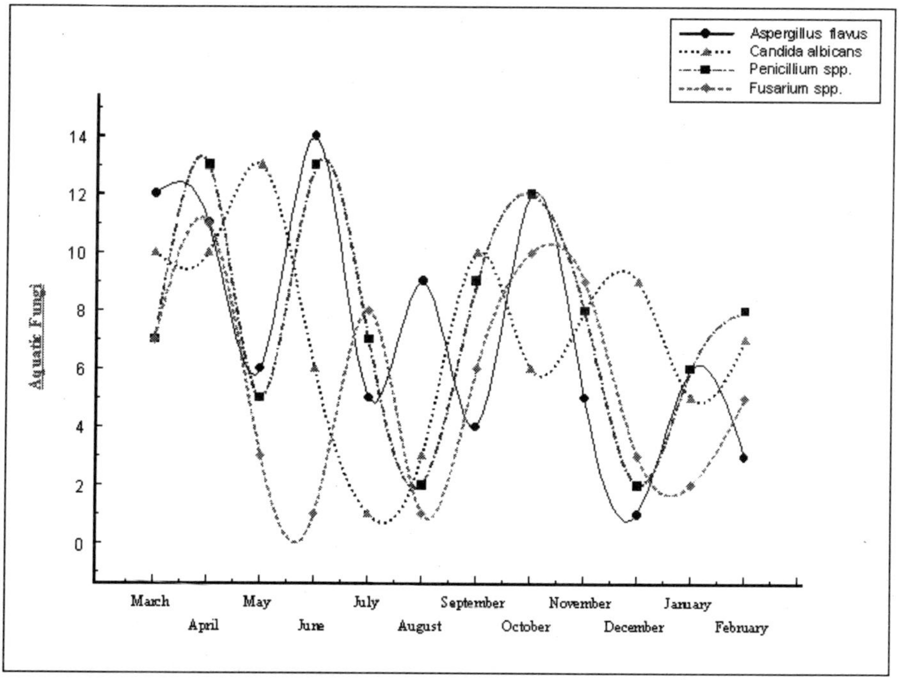

Figure 5.III.C.iii.11: Graph Showing the Monthly Variation in Aquatic Fungi at Sampling Site–IV (2008-2009)

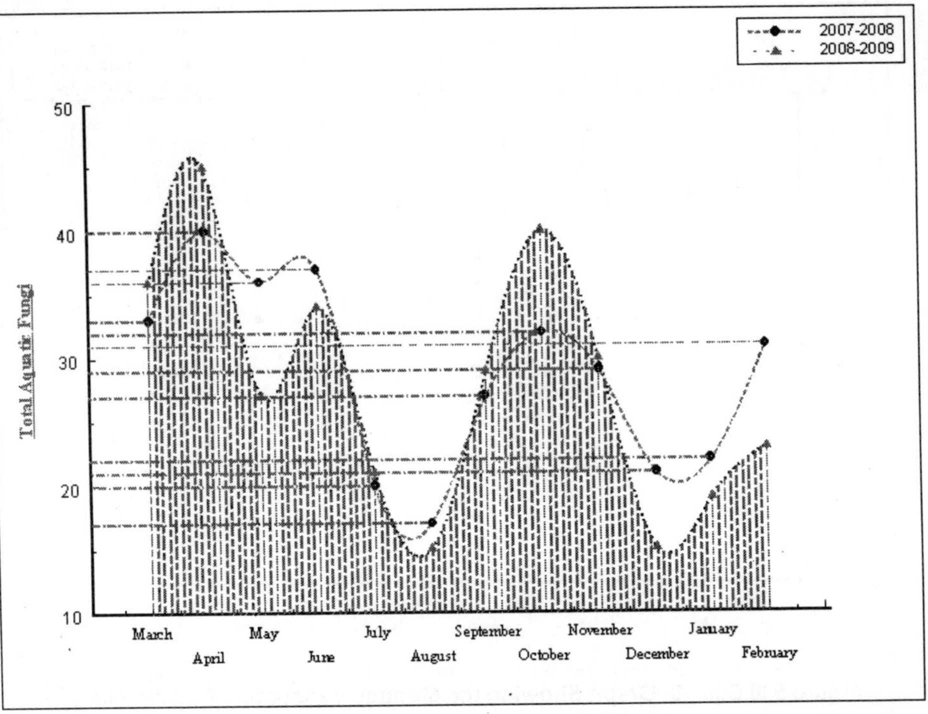

Figure 5.III.C.iii.12: Graph Showing the Monthly Variation in Total Aquatic Fungi at Sampling Site–IV (2007-2009)

Candida albicans in 2007-2008 was 7.17 $ml^{-1} x 10^3 \pm 3.93$ and 7.33 $ml^{-1} x 10^3 \pm 3.37$ in 2008-2009. *Penicillium* spp. showed optimum peak during the month of June in 2007-2008 (15 $ml^{-1} x 10^3$) and least peak during the month of August of 2007-2008 (00 $ml^{-1} x 10^3$). In 2007-2008 the average value of *Penicillium* spp. was 6.75 $ml^{-1} x 10^3 \pm 4.41$ and in 2008-2009 it was 7.67 $ml^{-1} x 10^3 \pm 3.73$. *Fusarium* spp. were found to be highest in the month of April 2008-2009 (11 $ml^{-1} x 10^3$) and recorded lowest in the same year in the month of June and August (01 $ml^{-1} x 10^3$). The average count of this species in 2007-2008 was 5.17 $ml^{-1} x 10^3 \pm 2.48$ and 5.50 $ml^{-1} x 10^3 \pm 3.53$ in 2008-2009.

Chapter 6
Discussion

Abiotic and biotic factors of an aquatic ecosystem are interdependent and the fluctuations of abiotic ones frequently affect the flora and fauna, changing their quality and diversity. Even though most of the bodies like ponds, lakes in earlier days were originally meant for other drinking purpose, but in today's life these water bodies are being utilized for various other purposes. The anthropogenic activities such as human interferences are disturbing the water quality of the ecosystem which has got specific morphoedaphic features.

The Paniyala Fish Pond is one of the major pond of the fisheries department of Roorkee city of District Haridwar. Being one of the major pond for the fish culture it was found necessary to assess the current limnological, microbiological and planktonic study of the pond.

Within the period of two years from March 2007 to February 2009, four physical, eight chemical and 6 biological parameters were investigated along with a study of fish diversity of the pond side by side.

Water temperature is one of the most important ecological factors which control physiological behaviour and distribution. Temperature of water depends upon the depth of water, solar radiation, climate and topography. Higher water temperature accelerates the rate of physiological reaction. Temperature also alters the saturation values of solids, and gases in water.

Das (1989) expressed that the lakes which do not come under oligotrophic and eutrophic status and fall intermediate between oligotrophic and eutrophic can be termed as mesotrophic. Sharma (1980) has also attempted to classify water bodies on the basis of water clarity values. Among all factors controlling the environment of pond water, temperature is one of the major factors having the most intense direct and indirect effect on the pond water on any aquatic ecosystem. In the present study the average temperature fluctuation ranged between maximum (29.70 °C) in summer season and minimum (17.63 °C) in winter season. The difference in water temperatures of the pond was less between summer and monsoon in comparison to summer and winter. However, the temperature showed a rising trend from end of winter to the middle of summer season.

Table: 6.1 Average Monthly Variation in Physical Parameters of Paniyala Fish Pond, 2007-2009

Months	*Water Temperature (°C)*	*Conductivity (µmhos/cm²)*	*Turbidity (JTU)*	*Total Solids (mg/l)*
		2007-2008		
March	25.85	39.75	48.00	222.64
April	27.85	38.00	78.25	237.39
May	29.60	40.50	82.00	251.29
June	29.60	40.00	101.00	436.78
July	29.15	41.00	135.80	799.32
August	27.50	51.75	140.80	735.52
September	28.83	56.00	36.00	723.49
October	29.33	82.25	19.75	262.07
November	21.48	93.50	18.75	118.50
December	17.63	81.75	22.25	88.30
January	19.45	73.25	29.25	95.31
February	22.45	59.25	31.50	94.28
Average ±SD	25.73 ±4.33	58.08±19.87	61.94±44.66	338.74±268.79
		2008-2009		
March	26.35	42.50	80.00	216.13
April	28.13	38.50	86.50	210.28
May	29.70	37.50	91.25	300.59
June	29.05	39.25	108.80	494.31
July	28.98	41.50	137.80	811.67
August	27.13	58.00	107.50	793.06
September	29.13	52.00	34.25	598.19
October	27.48	75.25	23.00	216.85
November	20.78	82.25	25.50	100.83
December	17.93	79.33	33.67	93.477
January	20.05	73.25	39.50	73.35
February	23.23	55.00	50.50	102.11
Average ±SD	25.66 ±4.09	56.19±17.17	68.18±38.59	334.24±271.76

SD ±: Standard Deviation.

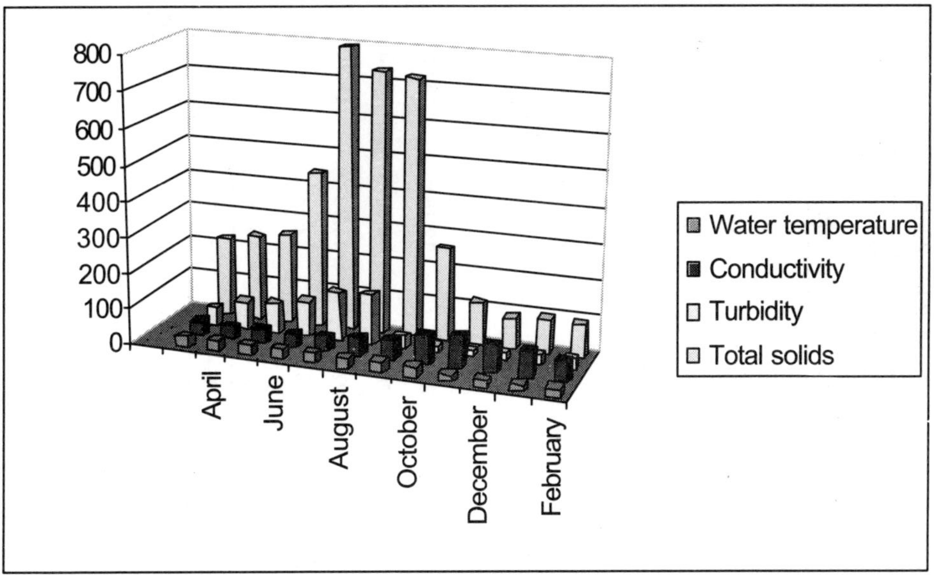

Figure 6.1: Average Value of Physical Parameters of Paniyala Pond during 2007-2008

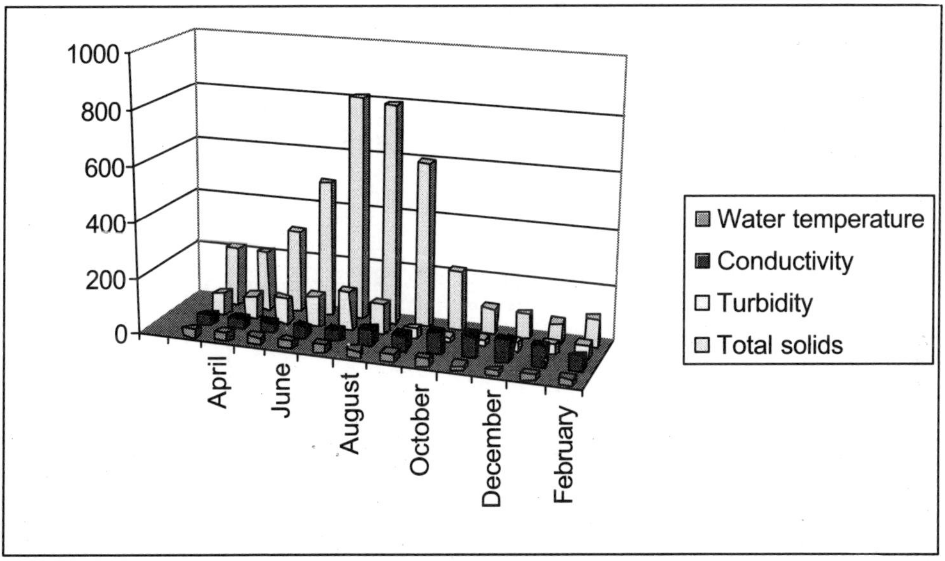

Figure 6.2: Average Value of Physical Parameters of Paniyala Pond during 2008-2009

Hutchinson (1957) reported that shallow water is influenced considerably by meteorological factors, such as air temperature and wind. During the present investigation, water temperature followed more or less similar trend as that of atmospheric temperature. These results are well in agreement with those of Kaur *et al.* (1995).

Table 6.2: Average Monthly Variation in Chemical Parameters of Paniyala Fish Pond, 2007-2009

Months	*pH*	*DO (mg/l)*	*BOD (mg/l)*	*COD (mg/l)*	*Free CO_2 (mg/l)*	*Total Alkalinity (mg/l)*	*Hardness (mg/l)*		*Chlorides (mg/l)*
							Ca	*Mg*	
				2007-2008					
March	8.24	6.70	3.75	10.65	2.253	339.00	31.40	16.79	24.45
April	8.27	5.58	3.93	10.98	2.765	330.00	33.70	17.60	26.77
May	8.31	5.13	4.28	11.23	3.245	306.00	36.03	23.90	28.37
June	8.28	5.38	4.70	11.68	2.603	262.00	37.33	26.03	31.94
July	7.88	5.50	4.48	11.10	2.463	253.00	29.45	11.23	25.15
August	7.80	5.53	4.20	10.38	2.59	341.00	27.58	11.30	22.14
September	8.17	6.13	3.90	9.825	2.055	352.00	27.83	11.60	20.48
October	8.26	6.15	3.08	9.75	2.258	425.00	29.08	12.18	18.68
November	8.23	8.35	2.98	9.425	1.28	347.00	28.00	13.05	16.80
December	8.23	8.93	2.83	9.65	1.178	341.00	27.83	14.20	17.25
January	8.21	8.13	3.08	9.975	1.398	322.00	27.63	14.70	19.30
February	8.22	7.60	3.33	10.43	1.643	338.00	28.50	15.76	21.78
Average±Sd	8.17±0.16	6.59±1.33	3.71±0.64	10.42±0.72	2.144±0.65	329.00±44.20	30.36±3.47	15.69±4.84	22.76±4.69

Contd...

Table 6.2–Contd...

Months	pH	DO (mg/l)	BOD (mg/l)	COD (mg/l)	Free CO_2 (mg/l)	Total Alkalinity (mg/l)	Hardness (mg/l)		Chlorides (mg/l)
							Ca	Mg	
				2008-2009					
March	8.32	6.63	3.73	10.90	2.813	338.00	29.55	16.79	23.58
April	8.37	5.55	4.05	11.18	3.63	335.00	32.98	18.05	25.64
May	8.34	4.98	4.28	11.40	3.225	309.00	36.10	24.98	28.25
June	8.28	5.13	4.73	11.73	2.588	273.00	37.25	26.05	32.30
July	7.84	5.50	4.50	11.48	1.915	264.00	29.35	12.30	28.95
August	7.73	5.70	4.18	10.65	2.738	291.00	27.43	11.93	23.33
September	8.12	6.43	3.85	10.18	1.375	328.00	27.88	12.03	22.05
October	8.27	6.48	3.28	9.475	1.218	385.00	28.78	12.73	18.85
November	8.24	8.43	3.13	8.725	1.365	364.00	28.33	11.08	17.15
December	8.23	8.90	2.93	8.975	2.215	351.00	27.50	14.25	17.13
January	8.20	8.25	3.18	9.825	2.915	324.00	27.43	14.53	19.05
February	8.22	7.70	3.45	10.43	2.788	338.00	26.90	16.28	21.30
Average±Sd	8.18±0.20	6.64±0.37	3.77±0.59	10.41±1.00	2.399±0.78	325.00±35.73	29.95±3.53	15.91±4.98	23.13±4.88

SD ±: Standard Deviation.

Welch (1952) has suggested that smaller the body, more quickly it reacts to the changes in atmospheric temperature. During rainy season, lowering of temperature was observed which might be due the cloudy weather and influx of rain water. Temperature remained low during the winter due to extreme cold and shorter sunshine period. Many researchers have shown that, temperature of water during winter are under the influence of high specific heat of water and overturn condition of ponds and other water bodies.

Conductivity determines the total amount of ionisable salts in water. It is due to ionization of dissolved inorganic solids. Seasonal fluctuations in this factor were caused mainly by variations in the ionic precipitations and diluting effect of rains (Welch, 1948).

Conductivity is an important factor for detecting fish community structure and plays a vital role in the productivity of aquatic ecosystem, also opinioned by Kaul *et al.* (1983). Conductivity of Paniyala Fish Pond was found to fluctuate between 37.5 to 93.50 μmhos/cm^2. Bhatt *et al.* (1999) observed the conductivity between 27 to 411 μmhos/cm^2 for Taudha lake. The maximum conductivity was observed in summer season. Maximum values of conductivity in summer might be due to water evaporation which resulted into increased electrolytes concentration in the pond water. During winter season, the water was steady and less disturbed and hence the salts must have settled down, resulting into lowering down the conductivity of water.

The average turbidity of pond water ranged between 18.75 JTU to 140.80 JTU, being maximum during monsoon and minimum during winter season. During monsoon season silt, clay and other suspended particles contribute to increase in turbidity. Seasonal variations in turbidity has also been reported by Ajmal and Razi-ud-din (1988). The turbidity of the water is actually the expression of optical property in which the light is scattered by the particles present in water. It is caused by the substances which are not present in the form of solution. Turbidity makes the water unfit for domestic purposes. Turbid water can adversely affect fish population by preventing successful development of fish eggs and larvae and by reducing the abundance of food available to fish.

Higher turbidity values were observed in monsoon at all the three stations, maximum at station II in both the years. Monsoon generally causes high turbulence and mixing of water leading to an increase in concentration of suspended particulate matter. In a study in similar lines Sakthivel and Shingdia (2001) observed rise in turbidity and total solids post on post Durga Pooja and idol immersion along silver beach in Mumbai. In summer and winter season, turbidity was found to be comparatively low. Bilgrami and Duttamunshi (1985) observed minimum values of turbidity in winter and summer seasons, while maximum in monsoon period.

The turbidity and total solids were observed to be closely and directly related to each other, as also reported by Verma *et al.* (1984) and Khanna (1993). The water is generally clean in other season with lowest turbidity in winters. The average total solid was maximum (811.67 mg/l) in monsoon season and minimum (73.35 mg/l) in winter season. The observations of Badola and Singh (1981) were also found to be similar. Dobriyal *et al.* (1983) also reported similar findings in the hill streams of

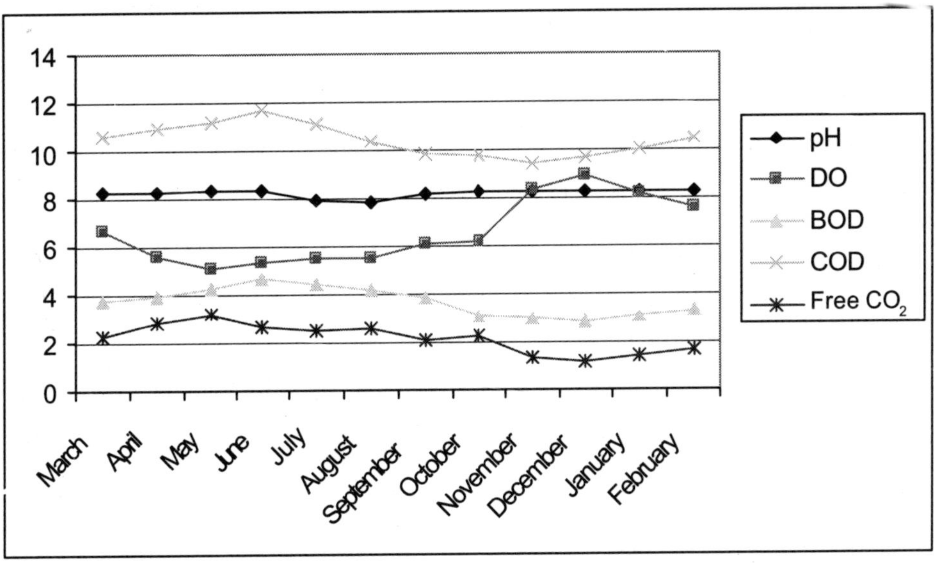

Figure 6.3: Average Value of pH, DO, BOD, COD and Free CO_2 of Paniyala Pond during 2007-2008

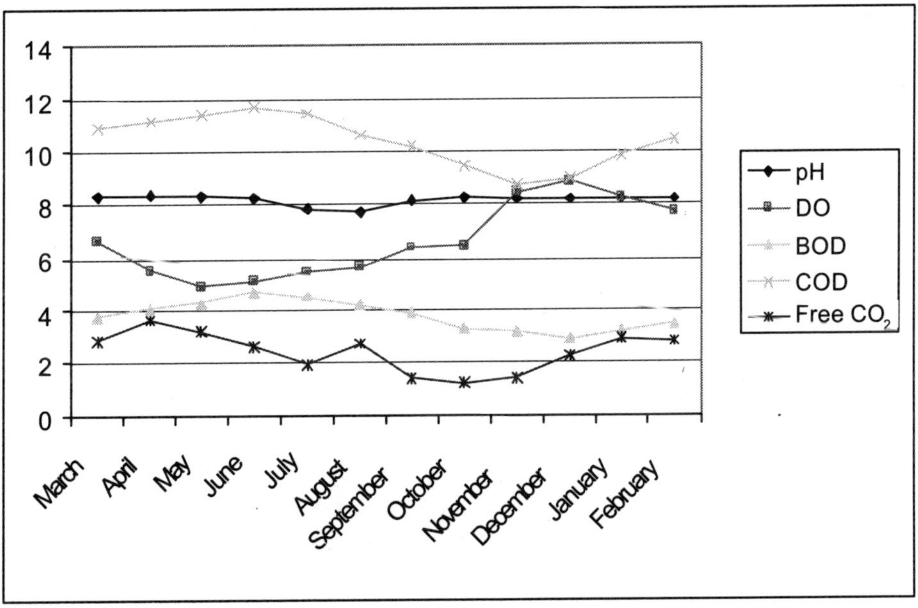

Figure 6.4: Average Value of pH, DO, BOD, COD and Free CO_2 of Paniyala Pond during 2008-2009

Garhwal. Singh *et al.* (1982) also observed this in the river Brahmaputra. The discharge of wastewater from near by industry may be the reason for this observation. Whether floating, coarse, fine, settelable or colloidal particles, the high load of total solids causes the water turbid and can influence on the survival of the aquatic living

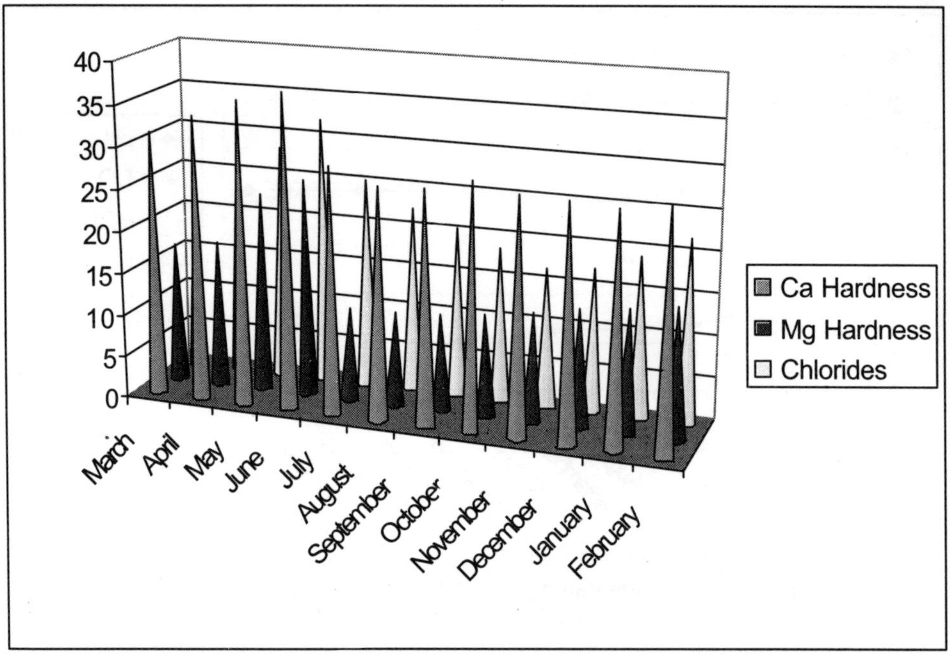

Figure 6.5: Average Value of Ca Hardness, Mg Hardness and Chlorides of Paniyala Pond during 2007-2008

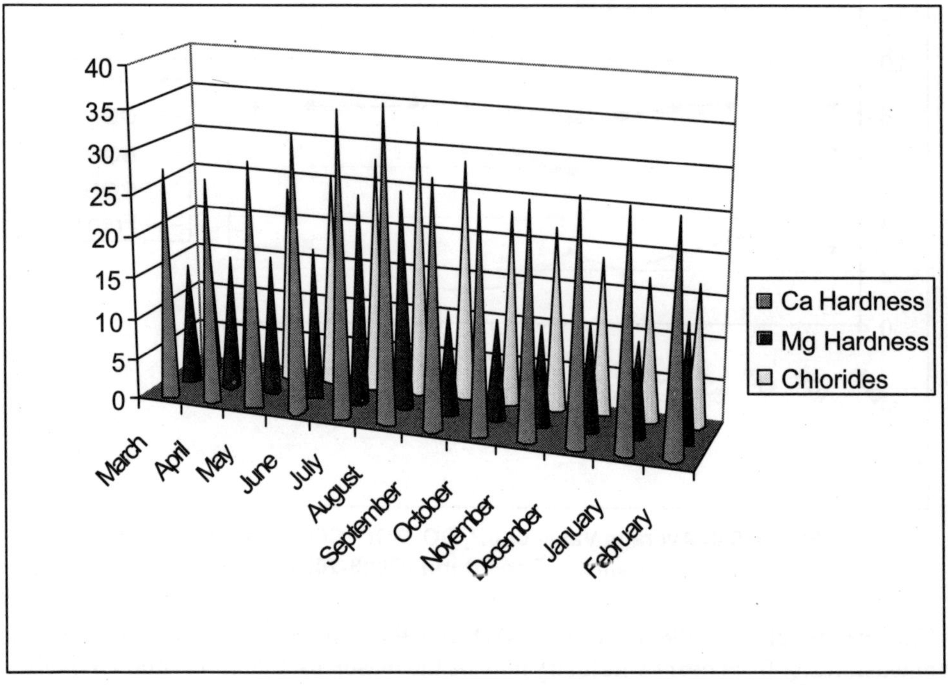

Figure 6.6: Average Value of Ca Hardness, Mg Hardness and Chlorides of Paniyala Pond during 2008-2009

Table 6.3: Average Quantitative Analysis of the Plankters of the Paniyala Pond During 2007-2009

Months	*Number of Plankters per litre of Water*				*Total Plankters (Unit/l)*	
	Phytoplankters (Unit/l)		*Zooplankters (Unit/l)*			
	2007-2008	*2008-2009*	*2007-2008*	*2008-2009*	*2007-2008*	*2008-2009*
March	2502.00	1447.25	460.75	313.25	2962.75	1760.50
April	2119.00	1490.00	413.75	302.25	2532.75	1792.25
May	1620.25	1291.25	245.00	195.50	1865.25	1486.75
June	854.25	893.75	91.50	86.25	945.75	980.00
July	595.25	611.00	39.75	45.25	635.25	656.25
August	771.75	765.75	54.00	55.25	825.75	821.00
September	1029.25	1081.50	104.75	95.75	1134.00	1177.25
October	1321.50	1483.25	141.25	180.25	1463.00	1663.50
November	1643.00	2170.50	216.25	227.75	1859.25	2398.25
December	2274.25	2757.00	301.50	296.50	2575.75	3053.50
January	3631.75	3269.50	411.75	425.50	4043.50	3695.00
February	3524.25	2509.25	385.50	366.25	3925.25	2875.50
Average±SD	1823.88±1019.10	1647.50±844.08	238.81±153.37	215.81±127.20	2064.02±1160.12	1863.31±956.98

SD ±: Standard Deviation.

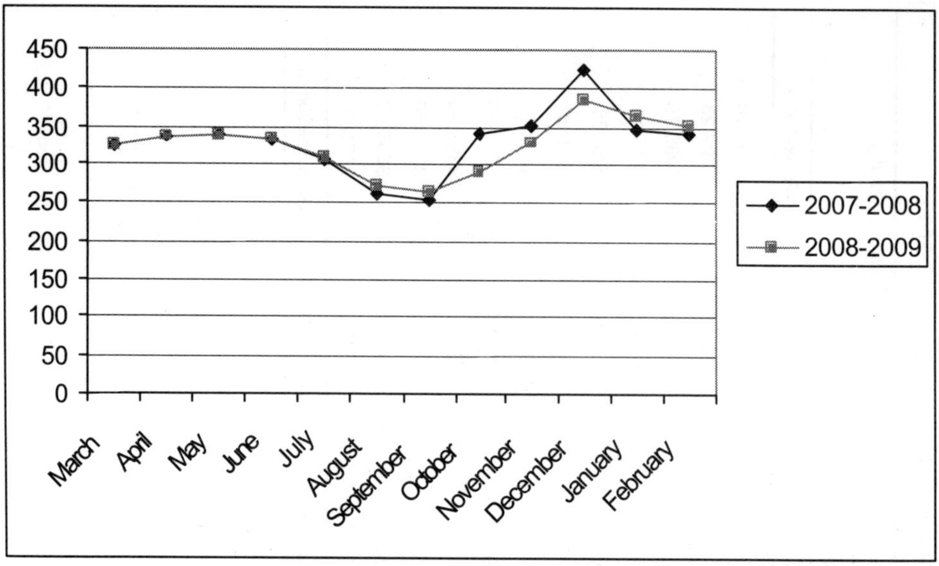

Figure 6.7: Average Value of Total Alkalinity of Paniyala Pond during 2007-2009

organisms. Further more, the mechanical abrasive action of the total solids may cause ecological imbalance. Penetration of light is limited to the suspended material; in the place where the water body has appreciable depth this may even restrict penetration of light. When turbidity is caused by the silt and clay it becomes a major limiting factor, but when it is due to living organisms it is an indices of productivity (Munshi and Munshi, 1995). The higher value of total solids also increased the turbidity in the river. Although many water bodies in India show the similar tendency in reference to fluctuation of turbidity as also observed by Verma and Shukla (1969), Gautam (1990), Badola and Singh (1981). Reddy and Venteshwarlu (1987) and Alabaster and Lloyd (1980) reported that concentration of suspended solids less than 25mg/l appeared to have effect on most fishes. Hynes (1970) reported that concentration of inert sediments that do not exceed 80 mg/l are thought unlikely to be of serious damage. Muller and Forstner (1968) also reported the general relationship between suspended sediment concentration and water.

Absolutely pure water is weakly dissociated and neutral in reaction *i.e.* H^+ or OH^- ions will be present in equal proportions. So since the dissociation constant remains the same, the concentration of opposite ions changes simultaneously. Limnologists in studies of fresh water systems have measured the pH. Hutchinson (1967) showed that pH in itself influences the natural distribution and abundance of aquatic animals, with the exception of few protozoan. It has somehow an indirect relation with natural life prevailing in water system. Fishes are relatively said to be poor regulators of internal pH, thus change in pH of their environment can alter enzyme activities or electrolyte composition of their body fluid, providing severe stress. Low pH also interferes with oxygen uptake, and pH outside range of 4.00 to 10.00 can kill fish (Mathew, 1998). The water of reservoir was found to be slightly alkaline during the course of study. The pH was maximum (8.3) in summer *i.e.* summer

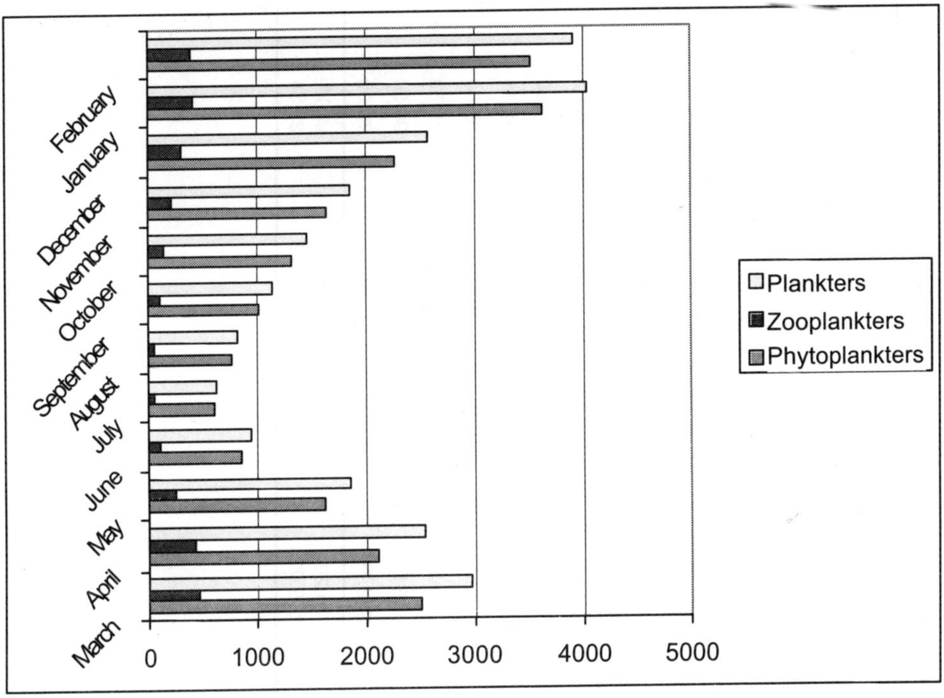

Figure 6.8: Average Value of Total Plankters of Paniyala Pond during 2007-2008

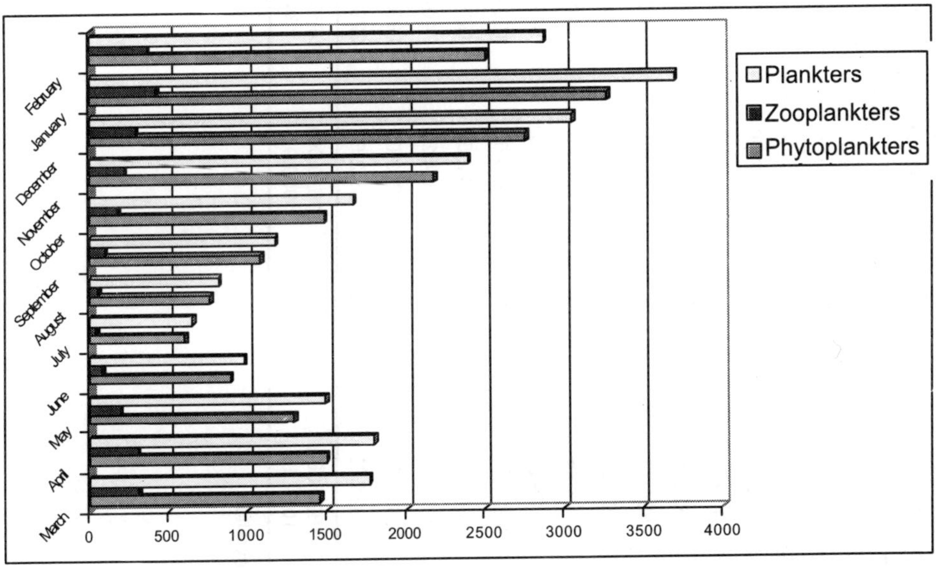

Figure 6.9: Average Value of Total Plankters of Paniyala Pond during 2008-2009

season. Higher concentration of calcium in water may also increase pH. The pH was recorded to be minimum (7.8) in July *i.e.* monsoon season. Similar trend have been reported by Badola (1979). The regression equation of pH in relation to Dissolved

Table 6.4: Average Number of different Group among the Phytoplankters of the Paniyala Pond During 2007-2009

Months	*Total Phytoplankters/ Liter of Water (Unit/l)*		*Total Diatoms/ Liter of Water (Unit/l)*		*Total Green Algae/ Liter of Water (Unit/l)*		*Total Blue Green Algae/ Liter of Water (Unit/l)*	
	2007-2008	*2008-2009*	*2007-2008*	*2008-2009*	*2007-2008*	*2008-2009*	*2007-2008*	*2008-2009*
March	2502.00	1447.25	2242.00	1197.75	221.25	213.50	38.25	55.00
April	2119.00	1490.00	1881.75	1236.75	177.00	187.00	59.50	73.00
May	1620.25	1291.25	1460.25	1141.25	123.75	124.25	35.50	32.00
June	854.25	893.75	764.50	780.25	71.00	99.00	18.25	15.75
July	595.25	611.00	556.50	556.75	35.00	48.50	3.00	7.00
August	771.75	765.75	781.75	714.00	34.25	45.50	5.25	6.00
September	1029.25	1081.50	891.00	947.75	100.75	108.00	22.00	25.50
October	1321.50	1483.25	1120.00	1298.75	165.50	145.50	35.25	38.75
November	1643.00	2170.50	1322.50	1920.50	286.75	228.50	35.75	35.00
December	2274.25	2757.00	1854.25	2407.00	350.50	296.75	72.00	70.50
January	3631.75	3269.50	3243.00	2833.25	318.00	362.75	70.50	96.50
February	3524.25	2509.25	3202.50	2231.25	255.75	221.50	65.75	73.00
Average±SD	1823.88± 1019.10	1647.50± 844.08	1610.00± 909.97	1438.77± 734.05	178.29± 109.02	173.40± 96.85	38.42± 24.13	44.00± 29.34

SD ±: Standard Deviation.

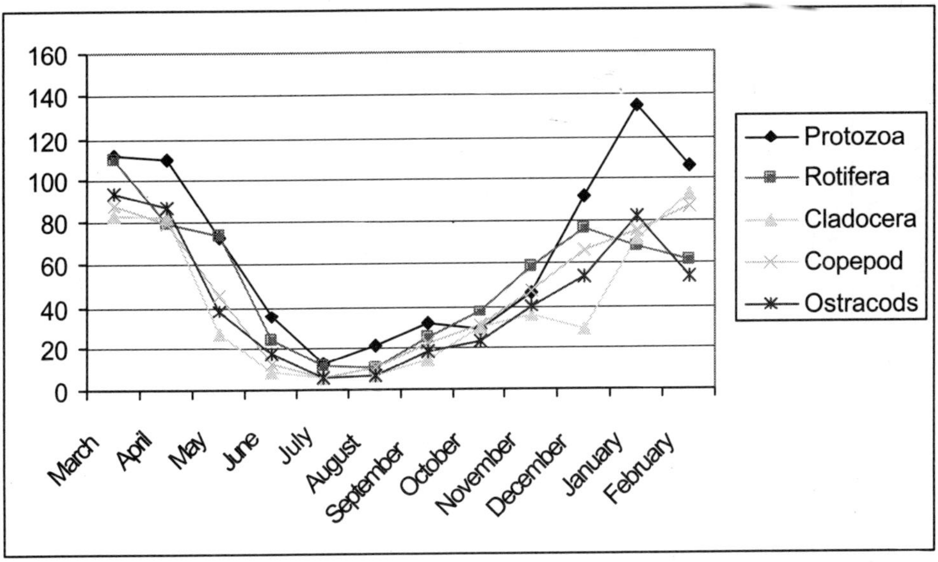

Figure 6.10: Average Value of Total Zooplankters of Paniyala Pond during 2007-2008

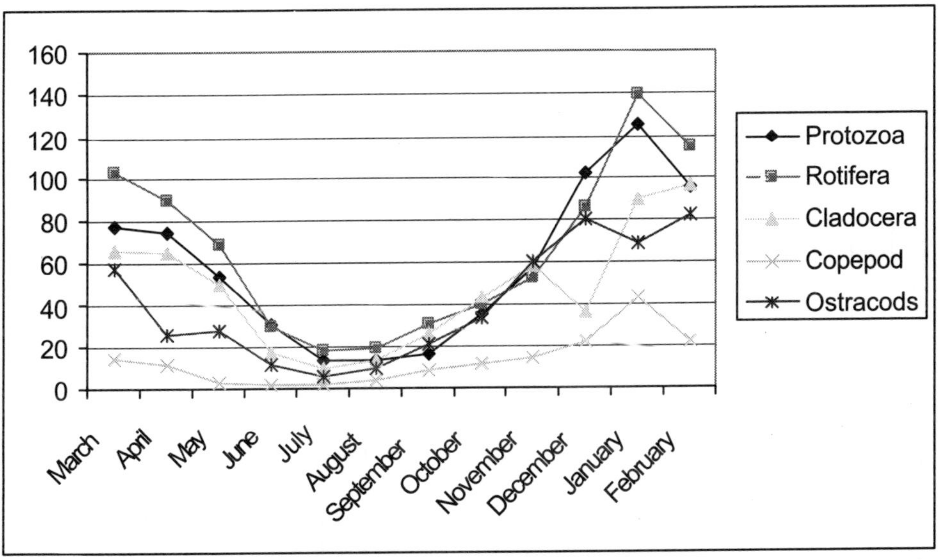

Figure 6.11: Average Value of Total Zooplankters of Paniyala Pond during 2008-2009

oxygen and Biochemical oxygen demand is shown as pH + 9.26 – 0.0526 DO – 0.196 BOD ($n = 12$, $df = 11_{(10,1)}$, $r^2 = 0.284$). However, in contrast to this, Khanna and Chugh (2004) observed higher pH value in winter season at Bhimgoda. Khanna (1993) also reported higher pH values in winter and less in rainy season.

Dissolved oxygen is one of the most important abiotic parameters influencing aquatic environment. Generally unpolluted water has higher DO and polluted water

Table 6.5: Average Number of Different Genera among Zooplankters in the Paniyala Pond During 2007-2009

Months	*Protozoa (Unit/l)*		*Rotifera (Unit/l)*		*Cladocera (Unit/l)*		*Copepod (Unit/l)*		*Ostracods (Unit/l)*	
	2007-2008	*2008-2009*	*2007-2008*	*2008-2009*	*2007-2008*	*2008-2009*	*2007-2008*	*2008-2009*	*2007-2008*	*2008-2009*
January	135.00	125.00	67.75	139.75	71.50	89.50	75.75	43	82.33	68.33
February	105.75	95.50	61.00	114.00	92.00	95.75	86.50	21.25	53.67	82.00
March	111.25	77.25	109.25	103.00	82.75	65.50	87.50	14.25	93.33	57.00
April	109.00	74.75	78.75	89.50	81.75	65.25	79.00	11.25	87.00	25.33
May	72.50	53.75	73.00	68.75	26.00	49.50	45.50	3.00	37.33	27.33
June	34.50	30.00	24.00	29.00	8.50	16.75	12.00	2.25	16.67	11.00
July	12.50	13.00	11.25	17.50	5.75	9.00	6.00	1.50	5.67	5.67
August	21.00	12.75	10.50	18.75	7.00	13.25	10.75	3.50	6.33	9.33
September	31.00	16.00	24.75	30.25	13.75	25.50	21.50	8.75	18.33	20.33
October	28.50	35.25	36.50	40.00	28.75	43.75	30.50	11.25	22.67	33.33
November	46.25	57.50	58.50	53.00	35.25	58.00	46.75	14.00	39.33	60.33
December	91.00	101.75	76.50	85.50	28.00	36.00	66.00	21.25	53.33	80.33
Average±SD	66.52±42.43	57.71±37.68	52.65±30.97	65.75±40.60	40.08±32.61	47.31±28.73	47.31±30.98	12.94±11.68	43.00±31.24	40.03±28.06

SD ±: Standard Deviation.

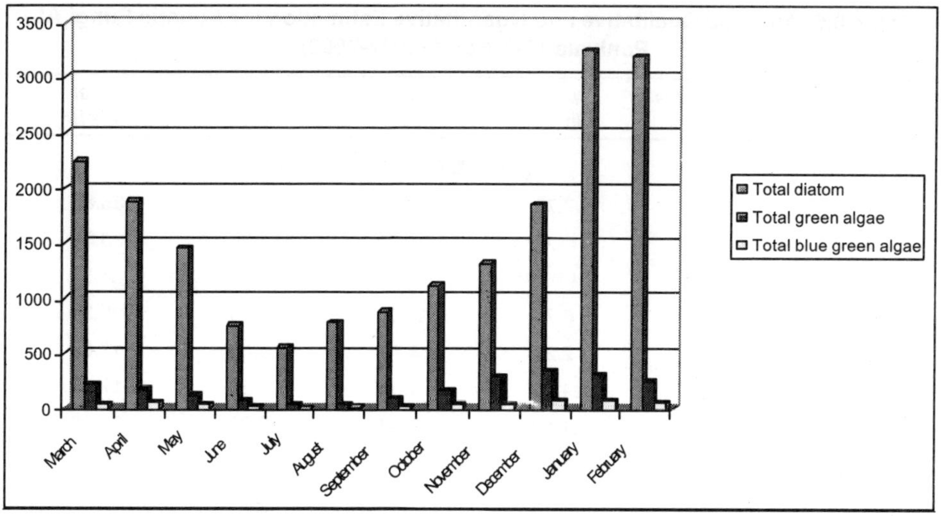

Figure 6.12: Average Value of Total Phytoplankters of Paniyala Pond during 2007-2008

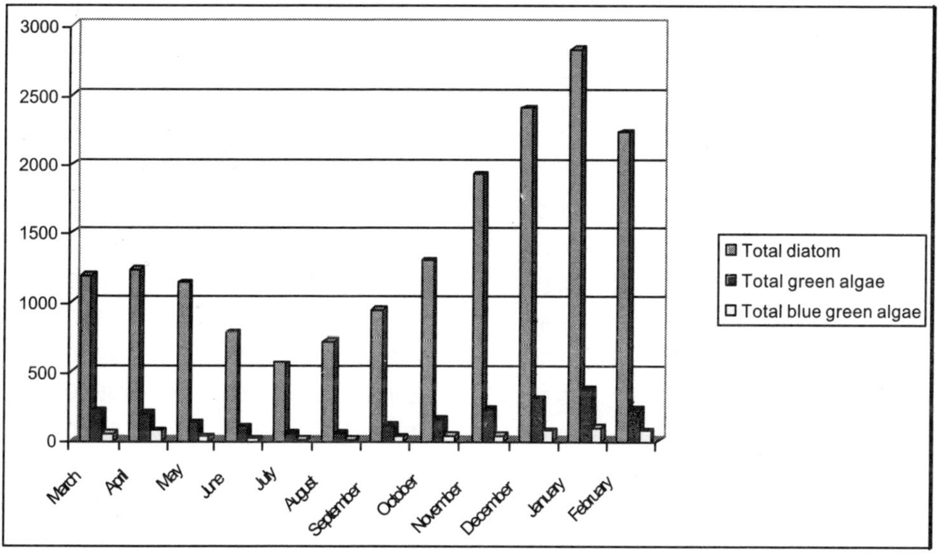

Figure 6.13: Average Value of Total Phytoplankters of Paniyala Pond during 2008-2009

has lower DO. There are two main sources of DO in an aquatic ecosystem (a) diffusion from air which depends on factors like wind action, temperature and salinity (b) photosynthesis which depends on transparency, turbidity, carbon dioxide and algal biomass. Depletion of DO to the level of anaerobia is the most critical manifestation of pollution and hence this factor can be used as an indicator for organic pollution. The maximum concentration of dissolved oxygen was recorded in winter, followed by

Table 6.6: Average Qualitative and Quantitative Estimation for Aquatic Fungi At Paniyala Fish Pond (2007-2009)

Months	*Aspergillus flavus*	*Candida albicans*	*Penicillium spp.*	*Fusarium spp.*	*Total ($ml^{-1}x10^3$)*
			(2007-2008)		
March	11.3	7.75	8.75	8.00	35.80
April	7.75	11.00	11.00	7.00	36.80
May	7.00	11.50	8.00	5.00	31.50
June	12.5	6.75	11.30	1.80	32.30
July	6.50	2.25	3.75	5.00	17.50
August	6.75	4.50	4.00	2.50	17.80
September	11.8	12.8	3.50	6.00	34.00
October	13.3	6.25	10.30	7.00	36.80
November	2.25	3.00	8.75	6.80	20.80
December	3.25	6.75	4.00	3.00	17.00
January	5.75	5.75	3.00	2.50	17.00
February	8.50	11.30	5.500	4.50	29.80
Average±Sd	8.04±3.54	7.46±3.47	6.81±3.17	4.92±2.10	27.20±8.49
			(2008-2009)		
March	10.00	11.50	8.00	7.00	36.50
April	10.30	11.50	10.30	7.80	39.80
May	7.75	10.00	7.25	5.80	30.80
June	11.8	4.75	9.75	2.30	28.50
July	4.75	2.50	6.50	6.80	20.50
August	6.00	5.50	2.25	2.00	15.80
September	8.50	7.50	5.75	5.80	27.50
October	12.80	5.50	9.75	7.50	35.50
November	2.75	4.75	9.50	6.30	23.30
December	2.00	7.00	4.25	3.00	16.30
January	7.75	5.50	5.50	3.50	22.30
February	5.00	4.25	6.75	4.30	20.30
Average±Sd	7.44±3.45	6.69±2.91	7.13±2.47	5.15±2.06	26.40±8.00

SD ±: Standard Deviation.

monsoon and minimum in summer season at the Paniyala pond. DO was found ranged between 5.0 mg/l to 8.93 mg/l during the period of investigation.

Lower oxygen content during summer season may be due to high temperature and decay of macrovegetation. Decomposition of organic matter goes on faster in warm than in cold season (Welch, 1948) which may result in depletion of oxygen in

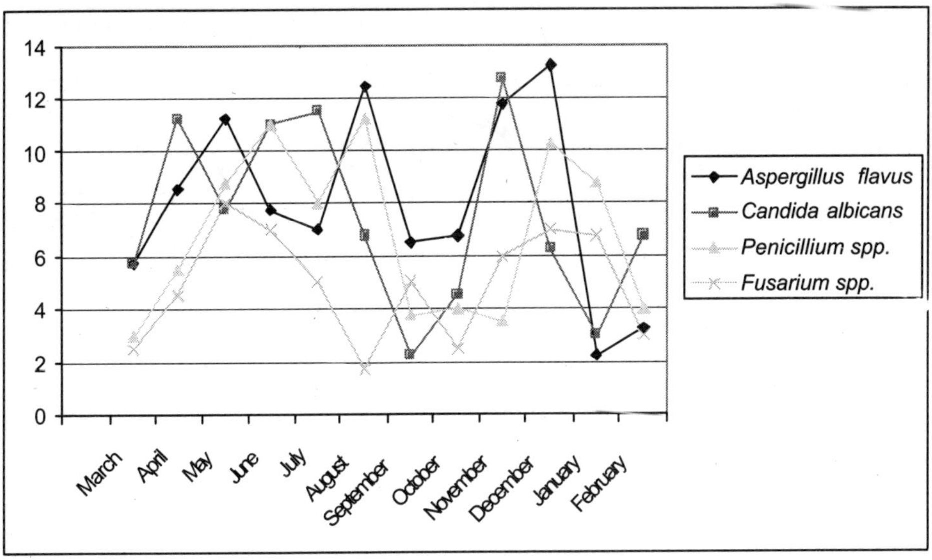

Figure 6.14: Average Value of Aquatic Fungi of Paniyala Pond during 2007-2008

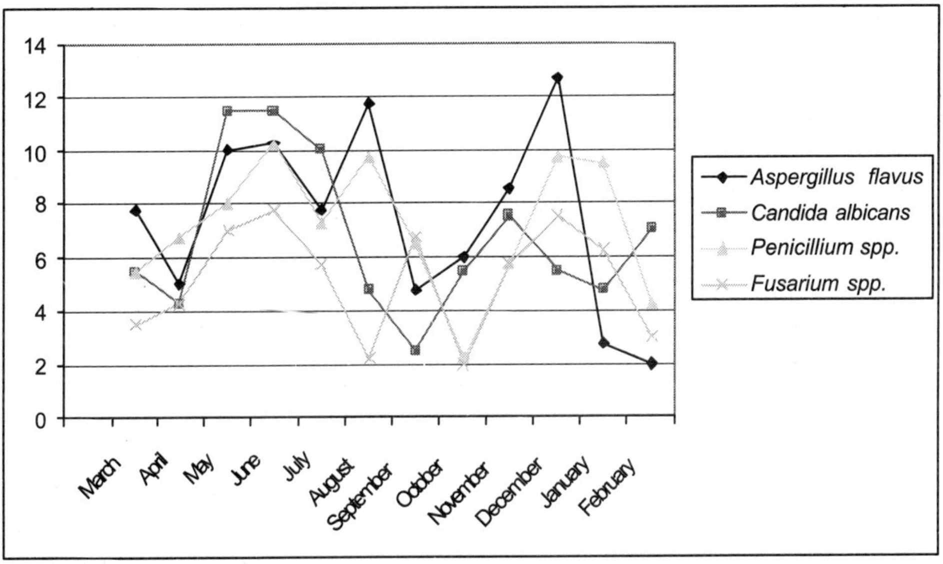

Figure 6.15: Average Value of Aquatic Fungi of Paniyala Pond during 2008-2009

summer. Singh (1965) also reported the high value of oxygen in winter due to the greater solubility of oxygen in water at lower temperature.

Kaushik and Saxena (1999) have reported high values of DO in monsoon season in the water bodies of central India and attributed it to circulating and ingressing rainwater into these water bodies. DO indicate a significant positive relation with transparency in the present investigation. At the same time, a significant inverse

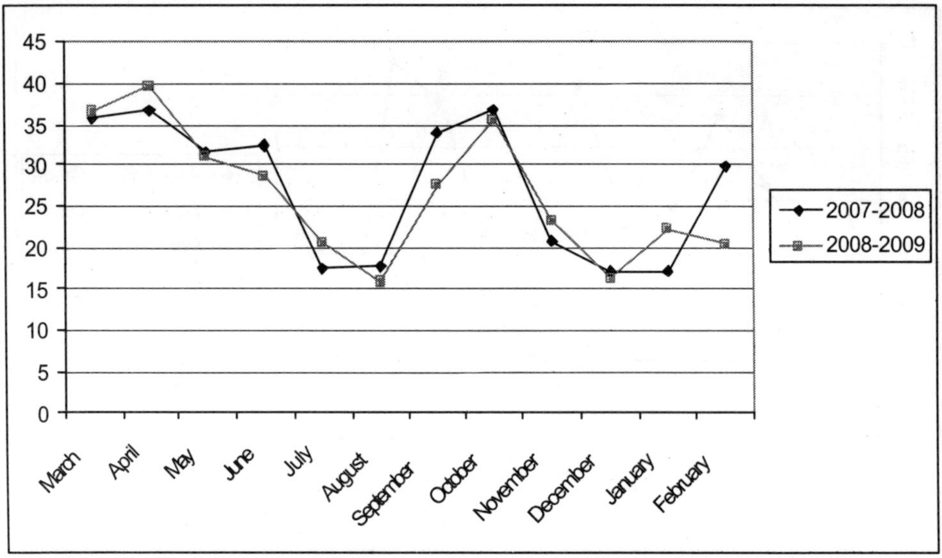

Figure 6.16: Average Value of Total Aquatic Fungi of Paniyala Pond during 2007-2009

relationship was observed between DO and Free CO_2. Parkash (1982) and Lohar and Bahura (1998) reported an inverse relationship of DO and temperature.

Table 6.7: Average Quantitative Analysis of Total Coliform Bacteria at Paniyala Fish Pond (2007-2009)

Months	*MPN/100ML*	
	2007-2008	*2008-2009*
March	44.75	49.75
April	59.50	59.00
May	66.50	71.50
June	86.75	82.00
July	123.25	107.75
August	149.25	148.00
September	166.75	176.50
October	133.50	155.50
November	14.75	13.25
December	10.25	9.00
January	8.00	11.25
February	11.25	14.50
Average ±SD	72.88±58.29	74.83±60.33

SD ±: Standard Deviation.

Table 6.8: Average Quantitative Analysis of Standard Plate Count at Paniyala Fish Pond (2007-2009)

Months	*SPC x 10^3*	
	2007-2008	*2008-2009*
March	42.25	51.00
April	57.50	58.75
May	60.75	69.50
June	79.50	81.25
July	118.25	119.25
August	162.75	154.50
September	173.5	167.00
October	158.75	147.25
November	19.25	41.50
December	10.00	11.00
January	11.25	11.00
February	13.25	16.00
Average ±SD	75.58±62.59	77.33±56.89

SD ±: Standard Deviation.

Low dissolved oxygen in summer in present investigation could be attributed to the high rate of oxidation of organic matter in water at with increase in temperature in which oxygen was consumed, secondly at higher temperature the water has a lesser oxygen holding capacity and surplus oxygen was lost to the atmosphere (Ruttner, 1953; Reid, 1961; Munawar, 1970). High metabolic rate in summer might be one of the important reasons for depletion of dissolved oxygen in summer. Similar situation was observed by Pawar and Mane (2006) at Sadatpur reservoir, Pravarnagar (M.S.).

All oxidizable organic material when degraded by aerobic microorganisms consumed the oxygen present in water. The oxygen consumed by organisms is proportional to the magnitude of organic matter present in water. BOD is lower in unpolluted water while it is high in case of polluted water (Hynes, 1960). The BOD is direct measure of biodegradable organic matter. It is an indicator parameter to know the presence of biodegradable matter in the wastes and express degree of contamination.

BOD showed a significant positive correlation with temperature. Rise in temperature increases the values of BOD as also have been reported by Gambhir (1999) in Maithon reservoir. A negative correlation has been observed between BOD and DO contents. A similar pattern has been reported by Khanna (1993), Chugh (2000). In the current study the regression equation was developed to study the relationship between the DO, BOD and COD [DO + 14.1 – 1.97 BOD – 0.007 COD ($n = 12$, $df = 11_{(10,1)}$, $r^2 = 0.649$)]. BOD in the current was found ranging between 2.83 mg/l to 4.73 mg/l during the period of investigation. Increased BOD in monsoon might be

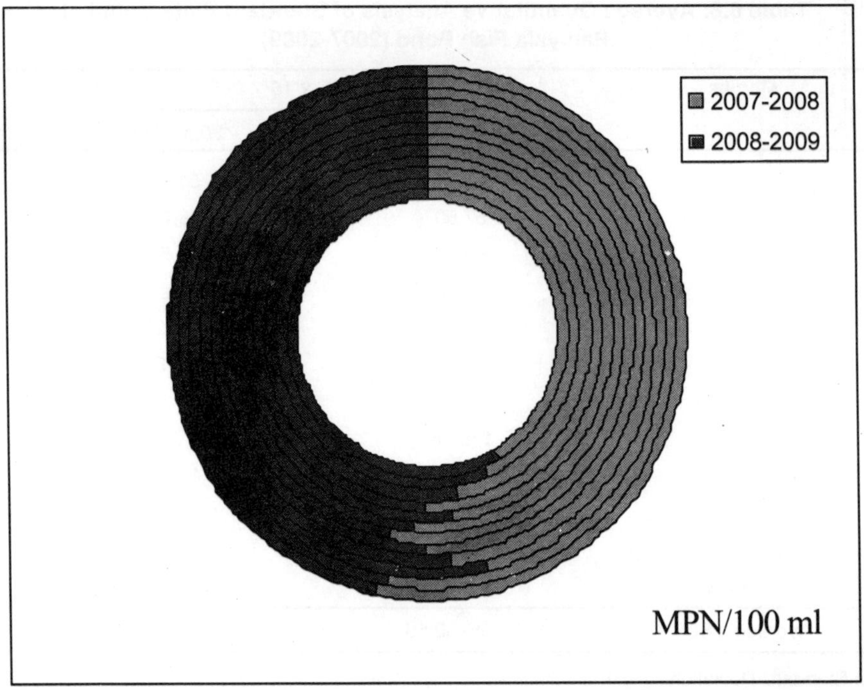

Figure 6.17: Average Value of MPN of Paniyala Pond during 2007-2009

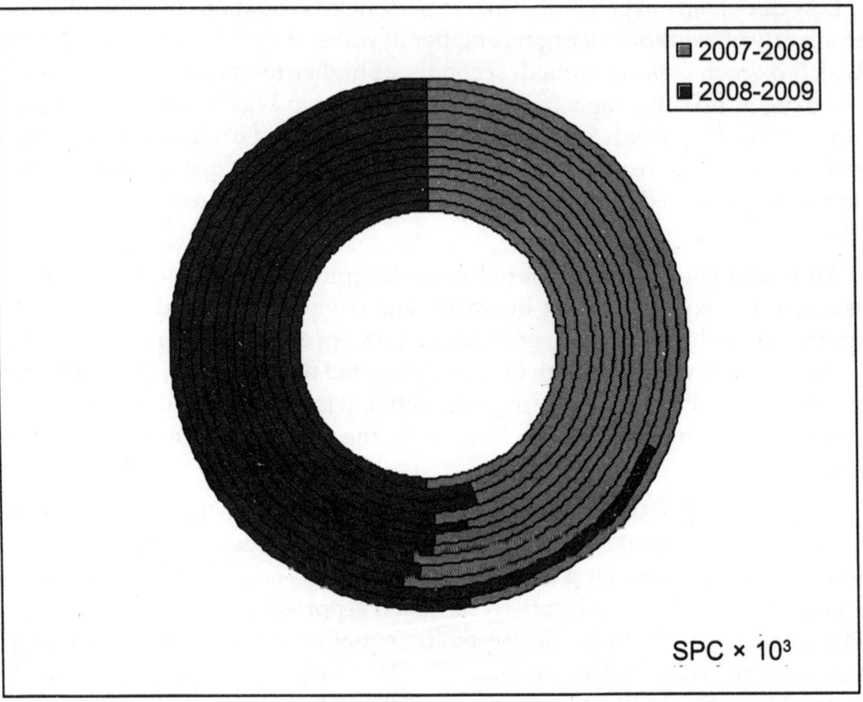

Figure 6.18: Average Value of SPC of Paniyala Pond during 2007-2009

Table 6.9: Statistical Analysis for Correlation of Physico–chemical Parameters

	Water Temp.	*Conducti–vity*	*Turbidity*	*Total Solids*	*pH*	*DO*	*BOD*	*COD*	*Free CO_2*	*Total Alkalinity*	*Ca Hard-ness*	*Mg Hard-ness*	*Chlorides*
Water Temp.		–0.724	0.597	0.657	–0.151	–0.960	0.816	0.699	0.507	–0.313	0.563	0.309	0.725
Conductivity			–0.746	–0.467	0.072	0.806	–0.873	–0.946	–0.811	0.653	–0.652	–0.552	–0.901
Turbidity				0.722	–0.578	–0.753	0.885	0.814	0.641	0.810	0.442	0.257	0.781
Total Solids					–0.777	–0.657	0.725	0.435	0.140	–0.536	0.035	–0.188	0.436
pH						0.207	–0.315	–0.077	0.067	0.377	0.433	0.544	–0.003
DO							–0.899	–0.818	–0.689	0.464	–0.652	–0.420	–0.829
BOD								0.909	0.655	–0.780	0.663	0.491	0.929
COD									0.819	–0.755	0.753	0.666	0.975
Free CO_2										–0.453	0.658	0.651	0.757
Total Alkalinity											–0.461	–0.420	–0.758
Ca Hardness												0.918	0.848
Mg Hardness													0.739
Chlorides													

due to discharge of domestic sewage and waste material especially during immersion activity. High BOD in monsoon was also reported by Patki Saroj S. (2002). Paliwan (2005) have reported low BOD values in summer. The lower BOD values in summer were also observed by Mishra *et al.* (1999). High BOD values in monsoon and winter was recorded by Bandela *et al.* (2002).

In chemically polluted water due to the presence of toxicants/ chemicals in which microbacterial degradation is not possible, BOD cannot be determined accurately. In such water degree of organic pollution can be assessed by chemical oxygen demand (COD). COD determines the amount of oxygen required for chemical oxidation of organic matter using a strong chemical oxidant. COD values, in present investigation were ranged between 8.725 mg/l to 11.73 mg/l. Maximum values of COD were recorded during monsoon followed by summer whereas minimum COD was found in winter. Salwe and Hiware (2006) also reported maximum COD in monsoon and summer and minimum in winter. Higher values in monsoon may be due to inflow of dead organic matter. Minimum COD in winter is due to settlement effect. Kudesia *et al.* (1986) also observed similar trend. Higher values of COD indicate the higher microbial activities and presence of oxidizable organic matter. Khulbe and Durgapal (1993) have also reported similar findings.

Carbon dioxide is of vital importance in the life of plant and microorganisms. It is produced as a result of respiration of aquatic organisms. As free CO_2 is highly soluble in water, it is found to be larger amount in polluted water as compared to fresh water bodies. CO_2 has a great effect on photosynthesis which affect fish growth. Surface area constantly absorbs or gives up carbon dioxide to maintain an equilibrium with the atmosphere. Carbon dioxide is an end product of both aerobic and anaerobic bacterial oxidation. Carbon dioxide is highly soluble in water. The respiratory activity of aquatic organisms and decompositicn of organic matter are important sources of carbon dioxide in fresh water body. Free CO_2 dissolved in water is the only source that can be used in photosynthetic activity of aquatic autotrophs. The daily and seasonal flow in the aquatic system forms the very base of food pyramid (Goldman and Horne, 1983).

Free CO_2 was recorded in all the months of study period with its maximum concentration in summer and minimum in winter season. High level of free CO_2 in summer and monsoon season could be coincided with the decline in the water level and increased load of organic matter. The concentration of free CO_2 usually does not exceed more than 20 ml/l (Doudoroff, 1957). However, it may reach as high as 50 mg/l in an organically polluted water body (Hynes, 1970). In the present study, the average range of CO_2 was found to be 1.17 mg/l to 3.24 mg/l. In the present study regression equation was developed between Free CO_2 and Alkalinity showing the inter-relationship Alkalinity + 394 – 29.2 Free CO_2 (n = 12, df = $11_{(9,2)}$, r^2 = 12.18)

Alkalinity of water is significant for biological activities. Alkalinity in water is its capacity to neutralize a strong base, is characterized by the presence of hydroxyl ions. As the acidic water is dangerous for fish growth, the ponds or reservoirs should have alkalinity so that the acid can be neutralized and fish production is possible.

The alkalinity is considered as an indicator of pond productivity (Davis, 1955). Alkalinity determines the availability of free CO_2 which is essential for photosynthesis and thus it is directly related to productivity. Thus, alkaline water is more productive and support the diversity of aquatic life.

In present investigation, the average total alkalinity of pond water was found in the range of 253.00 mg/l to 425.00 mg/l and the regression equation between pH and alkalinity was calculates as pH = 7.61 + 0.00174 Alkalinity (n = 12, df = 11 $_{(10,1)}$, r^2 = 0.174). Seasonal data showed maximum total alkalinity in summer, followed by winter and minimum in monsoon. According to Jhingran (1982), the range of total alkalinity in Indian water may be found from 40 mg/l to over 1000 mg/l. A well marked seasonal fluctuation in total alkalinity has been reported by majority of authors who have encountered minimum values in monsoon season and maximum values in summer season (Munawar, 1970; Ayyappan and Gupta, 1981).

Hardness of water is the sum of concentration of alkaline earth metal cations present in it. Calcium and magnesium are the principal cations imparting hardness. Calcium is an important micronutrient in an aquatic environment. Calcium is found in abundance in all natural waters and its source lies in the rocks from which it is leached. Calcium being an important contributor to hardness in water and reduces the utility of water for domestic use. It is an important constituent in exoskeleton arthropods and shells in molluscs (Pisks, 2000).

In the present study, average calcium contents of pond varied from 26.90 mg/l to 37.32 mg/l. Maximum calcium concentration was recorded in summer. Monsoon months also showed somewhat similar concentration of calcium. Increase of calcium hardness during summer may be attributed to evaporation of surface water.

Basin geology has direct impact on calcium hardness. In urban environment, two other anthropological impacts are worth mentioning, one is construction activities and second idol immersion (Kodarkar and Chandrashekkar, 1995). Lund (1964) suggested calcium has main effect on phytoplankters by buffering pH of water. In the present investigation, a feeble negative relation was observed between calcium and phytoplankters except group Myxophyceae.

The ecological significance of major cations *i.e.* calcium and magnesium in the biotic dynamics of aquatic fauna and flora is well established fact. Magnesium is essential for flora for chlorophyll biosynthesis and enzymatic transformation (Wetzel, 1975). Magnesium forms the nucleus of the porphyrine ring of the chlorophyll and therefore is an essential constituent. During the present study a regression equation was developed to show the relationship between Chlorides, Ca and Mg Hardness [Chlorides = - 12.4 + 1.46 Ca Hardness – 0.242 Mg Hardness (n = 12, df = 11 $_{(10,1)}$, r^2 = 1.731)]. The demand for magnesium in metabolism is minor as comparison to the quantities generally available in fresh water. In present investigation, the average quantity of magnesium of pond water was found in the range of 11.08 mg/l to 26.05 mg/l and the average chloride of pond water was found in the range of 16.8 mg/l to 32.3 mg/l. Magnesium showed a significant positive relation with chloride as compared to calcium and total alkalinity.

Phytoplankters are the pioneer of an aquatic food chain. The productivity of an aquatic environment is directly correlated with the density of phytoplankter. The phytoplankter is the base of most of the lake food web and fish production is linked to phytoplankter. Number and species of phytoplankton serves to determine the quality of water body. Phytoplankters form a bulk of food for zooplankter, fishes and other aquatic organisms. In present investigation, the average number of phytoplankter of pond water was found in the range of 595.25 to 3631.75 in per litre of water.

The maintenance of a healthy aquatic ecosystem depends on the abiotic properties of water and biological diversity and is the very useful tool for assessment of water quality. Kalyani *et al.* (1999) have recorded bimodal peak of phytoplankton population *i.e.* summer peak in April and winter peak in December. Similar peak in phytoplankters population were observed by Das and Shrivastava (1959). Bhowmic (1968) reported maximum number of phytoplankton during monsoon and winter. While Michael (1969) has reported a prolonged single peak. The effect of temperature on phytoplankton can not be separated from the effect of light since both the factors are interrelated in photosynthesis. The enhanced growth of phytoplankton in summer could be attributed to increase temperature and light during summer [Temperature = 11.2 – 0.0375 Phytoplankters (n = 12, df = 11 $_{(9,2)}$, r^2 = 1.619)].

Zooplankters are important index of secondary production and a natural source of food for higher organism, including fishes. Zooplankter plays a key role in transferring energy from one trophic level to other in aquatic habitats. The zooplankters play a significant role in an aquatic system as consumers. In present investigation, the average number of zooplankters of pond water was found in the range of 45.25 to 460.75 in per litre of water. In relation to temperature the regression equation was Temperature = 9.8 – 0.0179 zooplankters (n = 12, df = 11 $_{(10,1)}$, r^2 = 1.545).

The abundance of zooplankters has been governed by cumulative effects of physico-chemical variables. Singh (1986) reported fluctuation in the composition of zooplankton population in relation to hydrobiological conditions.

Prasad (1956) has stated that the temperature is the determining factor in the seasonal distribution of organisms. In the present study, a very feeble positive relation was observed between zooplankters and temperature.

Welch (1952) mentioned that, the fluctuation in zooplankton population is greatly influenced by the variation in temperature along with many other factors. Among the several factors, temperature seems to exhibit the greatest influence on the periodicity of zooplankton (Battish and Kumari, 1996). However, in shallow tropical perennial and seasonal ponds, such a regular food cycle cannot be seen. According to Reynolds *et al.* (1968) pH is considered to be the most important factor regulating the plankton density. In the present study, pH was alkaline and showed a significant positive relation with total zooplankters.

By the analysis for correlation coefficient (r) parameters, it is extracted that there are highly significant correlation between Water Temperature – Conductivity (-0.724), Water Temperature – Turbidity (0.597), Turbidity – Conductivity (-0.746), Water Temperature – Total Solids (0.657), Conductivity – Total Solids (-0.467), Total Solids

– Turbidity (0.722), Water Temperature – pH (-0.151), pH – Conductivity (0.072), pH – Turbidity (-0.578), pH – Total Solids (-0.777),Water Temperature – Dissolved Oxygen (-0.960), Conductivity – Dissolved Oxygen (0.806), Turbidity – Dissolved Oxygen (-0.753), Total Solid – Dissolved Oxygen (-0.657), pH – Dissolved Oxygen (0.207), Water Temperature – BOD (0.816), Conductivity – BOD (-0.873), Turbidity – BOD (0.885), Total Solid – BOD (0.725), pH – BOD (-0.315), Dissolved Oxygen – BOD (-0.899), Water Temperature – COD (0.699), Conductivity – COD (-0.946), Turbidity – COD (0.814), Total Solid – COD (0.435), pH – COD (-0.077), Dissolved Oxygen – COD (-0.818), BOD -COD (0.909),Water Temperature – Free CO_2 (0.507), Conductivity – Free CO_2 (-0.811), Turbidity – Free CO_2 (0.641), Total Solid – Free CO_2 (0.140), pH – Free CO_2 (0.067), Dissolved Oxygen – Free CO_2 (-0.689), BOD - Free CO_2 (0.655), COD – Free CO_2 (0.819), Water Temperature – Total Alkalinity (-0.313), Conductivity – Total Alkalinity (0.653), Turbidity – Total Alkalinity (0.810), Total Solid – Total Alkalinity (-0.536), pH – Total Alkalinity (0.377), Dissolved Oxygen – Total Alkalinity (0.464), BOD - Total Alkalinity (-0.780), COD – Total Alkalinity (-0.755), Free CO_2 – Total Alkalinity (-0.453), Water Temperature – Calcium Hardness (0.563), Conductivity – Calcium Hardness (-0.652), Turbidity – Calcium Hardness (0.442), Total Solid – Calcium Hardness (0.035), pH – Calcium Hardness (0.433), Dissolved Oxygen – Calcium Hardness (-0.652), BOD – Calcium Hardness (0.663), COD – Calcium Hardness (0.753), Free CO_2 – Calcium Hardness (0.658),Total Alkalinity – Calcium Hardness (-0.461),Water Temperature – Magnesium Hardness (0.309), Conductivity – Magnesium Hardness (-0.552), Turbidity – Magnesium Hardness (0.257), Total Solids – Magnesium Hardness (-0.188), pH – Magnesium Hardness (0.544), Dissolved Oxygen – Magnesium Hardness (-0.420), BOD – Magnesium Hardness (0.491), COD – Magnesium Hardness (0.666), Free CO_2 – Magnesium Hardness (0.651),Total Alkalinity – Magnesium Hardness (-0.420), Calcium Hardness – Magnesium Hardness (0.918), Water Temperature – Chlorides (0.725), Conductivity – Chlorides (-0.901), Turbidity – Chlorides (0.781), Total Solids – Chlorides (0.436), pH – Chlorides (-0.003), Dissolved Oxygen – Chlorides (0.829), BOD – Chlorides (0.929), COD – Chlorides (0.975), Free CO_2 – Chlorides (0.757),Total Alkalinity – Chlorides (-0.758), Calcium Hardness – Chlorides (0.848), Magnesium Hardness – Chlorides (0.739).

The analysis for correlation of the total plankters community (Phytoplankters and Zooplankters) is 0.897.

In natural ecosystem, macrophytes remove both toxic and non-toxic elements in the sediments and water (Narayana and Somashekar, 1997). Moreover, metabolic activities of macrophytic communities accelerate the physico-chemical conditions of the stream. Submerged aquatic macrophytes are one of the most important components of the littoral zones of lakes. Field studies conducted by Carpenter and Lodge (1986), Balls *et al.* (1989) and Sondergaard *et al.* (1996) evidenced that submerged macrophytes have an important role in restoration of shallow eutrophic lakes. Therefore, from an engineering point of view, evaluation of life cycles of macrophytes is a key issue before using them in restoration of lakes (Carpenter and Adams, 1977; Asaeda and Bon, 1997).

During the study some fishes were also observed. The fishes observed were *Catla-catla, Labeo-rohita, Cirrhinus-mrigala, Hypophthalmichthys-molitrix, Cyprinus-carpio, Channa-stratius, Notopterus-notopterus, Chanda-nama.*

Macrophytes also take up a large amount of nutrients from the sediment and water. But after death and decay, the nutrient load is again increased in the sediment. In the succeeding decomposition process, accumulated organic matter and nutrients are released, which has a significant influence on the lake ecosystem. The release of ammonia and phosphorous raises the nutrient concentration in the overlying waters and the oxygen consumption in the chemical and biological decomposition process may decrease the dissolved oxygen level in the overlying waters as well as sediments (Jewell, 1971; Pereira *et al.*, 1994).

During the present study total 14 different species of macrophytes were identified, belonging to 3 classes 14 families and 14 genera. Aquatic macrophytes are important indicators of water quality. Ponds harboured a luxuriant growth of aquatic plants. The prolific growth of submerged and free floating macrophytes add high amount of dissolved oxygen, particularly in winter as observed by Sahai and Sinha, (1969).

The present study reveals that the value of Most Probable Number (MPN) coliform bacteria increases from summer season, average maximum value was noted 176.50 MPN/ 100 ml and average minimum was observed 8 MPN/ 100 ml. Further in SPC the minimum average value were 10 x 10^3 SPC/ml and maximum average of the pond was 173.50 x 10^3 SPC/ ml.

While extending the study some nutrient parameters were also studied which play a major role in the growth and development of pond and its ecosystems. Almost all of the phosphorus (P) in water is in the form of phosphate (PO_4). Phosphorus is an essential plant nutrient, and because it is often in limited supply, adding phosphorus to water will stimulate plant (algae) growth. This growth of algae can be undesirable, as in the case of pristine streams, or desirable, as in ponds for fish culture. Considerable care in sample collection, handling and preservation is required to obtain truly accurate phosphorus values. In the current study the value recorded for the phosphate was 0.071 mg/l. Nitrate levels in drinking water for humans and livestock are a major concern. Typical levels in surface waters range from 0.005 to 0.5 mg/l nitrate. However, nitrate is relatively nontoxic to fish and is not a health hazard except at exceedingly high levels. During the current study the water sample was analysed and value of nitrate recorded was 0.23 mg/l for the Paniyala State Fish Pond. Rao (1987) opines that high TDS and bicarbonate values are well indicative of high nutrient loads. Dissolved solid in water play a very important role in the health of aquatic ecosystem, as if the amount of dissolved solid increases in water, the water quality is affected. In the present study the TDS value was found 42.15 mg/l.

However the present study of Paniyala State Fish Pond indicates low transparency, lower chloride content, moderate DO, low BOD and low values of nutrient which do not support rich productivity of water body. The pond water parameters have crossed the desirable drinking water criteria with respect to pH, TDS, total hardness, magnesium hardness etc. and hence it should be used with almost care by treating it to minimize the excess values of the above parameters, and to bring them below the desirable limits of drinking water standards.

Chapter 7
Summary

This thesis embodies the findings of study conducted under title "Microbial and Planktonic Ecology of Paniyala State Fish Pond at Roorkee, district Haridwar (Uttaranchal)" during March 2007 to February 2009. In this study several aspects of the ecological factors which affect the water system through direct or indirect ways, are analyzed and evaluated in various aspects.

The selection of Paniyala State Fish Pond at Roorkee for the study was done because of its economic importance to the district as well as to the state.

The results which were observed during this study are summarized as below:

- ☆ In the present study the average temperature fluctuation ranged between maximum (29.70 °C) in summer season and minimum (17.63 °C) in winter season.
- ☆ Conductivity means the ion present in the water responsible for the conduction of electricity, more the value of conductivity means more ions present in water. The conductivity of Paniyala Fish Pond was found to fluctuate between 37.5 μmhos/cm^2 to 93.50 μmhos/cm^2.
- ☆ Turbidity is one of the most important variables pertaining to the optical characteristics of the pond water. The average turbidity of pond water ranged between 18.75 JTU to 140.80 JTU, being maximum during monsoon and minimum during winter season.
- ☆ pH represents the number of hydrogen present in the water. The value less than 7.00 represents the acidic water and more than 7.00 represents the alkaline water status. The pH was maximum (8.3) in summer *i.e.* summer season and minimum pH was (7.8) observed in July *i.e.* monsoon season.

- Dissolved oxygen (DO) is one of the most important chemical factors having a widespread effect upon the flora and fauna of any aquatic ecosystem. It also tells us about the phytoplanktonic status of the water. In this study DO ranged between 5.0 mg/l to 8.93 mg/l.

- One of the well known water quality parameters BOD is very much responsible for the consumption of the DO in an aquatic ecosystem and for the degradation of water quality. BOD was found ranged between 2.83 mg/l to 4.73 mg/l during the period of investigation.

- The chemical oxygen demand is used as a measure of the oxygen equivalent of the organic matter content of a system that is susceptible to oxidation by a strong chemical oxidant. COD values, in present investigation were ranged between 8.725 mg/l to 11.73 mg/l.

- The CO_2 presents in water is used by the phytoplankters to photosynthesize. But more CO_2 may cause water to acidic. In the present study, the average range of CO_2 was found to be 1.17 mg/l to 3.24 mg/l.

- In present investigation, the average total alkalinity of pond water was found in the range of 253.00 mg/l to 425.00 mg/l.

- Hardness was understood to be a measure of the capacity of water to precipitate the soap. In the present study, average calcium contents of pond varied from 26.90 mg/l to 37.32 mg/l.

- The average quantity of magnesium of pond water was found in the range of 11.08 mg/l to 26.05 mg/l.

- Chloride, in the form of chloride ion (Cl^-) is one of the major inorganic anions in water system. Micro as well as higher organisms present in the water system for nutrition uses this anion in the trace. The fluctuation in chloride content of pond water was found in the range of 16.8 mg/l to 32.3 mg/l.

- The number of phytoplankter in any aquatic system are indicator of health of that system they also tell about the eutrophication status of the pond. In this study the average number of phytoplankter of pond water was found in the range of 595.25 to 3631.75 per litre of pond water.

- The present study comes out with a result of total 14 different species of macrophytes, belonging to 3 classes 14 families and 14 genera.

- The average value for coliform bacteria in Paniyala Pond was noted 8 MPN/ 100 ml to 176.50 MPN/ 100 ml.

- Further in SPC the minimum average value were 10 x 10^3 SPC/ml and maximum average was 173.50 x 10^3 SPC/ ml.

Concluding Remark

Present study reveals that Paniyala State Fish Pond is one of the major pond of the fisheries department of Roorkee city, District Haridwar. Being one of the major pond for the fish culture it is necessary to assess the current limnological, microbiological and planktonic study to maintain the water quality of the pond and government department should take necessary action to protect the pond from any contamination.

References

Ahemed, M. and Alireza, S., (1992): *Diurnal fluctuations in population density of zooplankton from Nathsagar, Paithan. Maharashtra Proc. Acad. Environ. Bio.*, 1 (20): 173-178.

Ajmal, M. and Razi-ud-Din, (1988): *Studies on the pollution of Hindon river and Kalinadi (India).* In Ecology and pollution of Indian rivers-(R.K. Trivedi Ed.) Asian Publishing House, New Delhi. pp: 87-111.

Alabaster, J.S. and Lloyd, R., (1980): *Water quality criteria for fresh water fishes,* Buttorworth, London.

Allan, J.D., (1995): *Stream Ecology: Structure and Functioning of Running Water,* Chapman and Hall, London.

Allan, J.D. and Flecker, A.S., (1993): Biodiversity and conservation of running water. *Bioscience,* 43: 32-43.

APHA, AWWA, WPCF, (1998): *Standard methods for the examination of water and wastewater,* 20th ed., Washington D.C., New York.

Arjariya, Amita, (2003): *Physico-chemical profile and plankters diversity of Ranital lake, Chhatarpur, M.P.,* J. Nat. Env. and Poll. Tech., 2(3): 327-328.

Aseada, T, and Bon, T.V., (1997): Modeling the effects of macrophytes on algal blooming in eutrophic shallow lakes. *Ecol. Modeling,* 104: 261-287.

Ayyappan, S. and Gupta, T.R.C., (1981): Limnology of Ramasamudra tank Hydrography, Mysore. *J. of Agri. Sci.,* 15: 305-312.

Bachmann, R.W., (2001): *The Limiting Factor Concept: What Stops Growth? Lakeline,* Spring. pp: 26-28.

Badola, S.P., (1979): *Ecological studies on the Ichythofauna of some freshwater resources of Garhwal Region*, D.Phil Thesis, Garhwal University

Badola, S.P. and Badola, Smita, (1999): *Observations on destructive fishing methods and recommendations to protect fish fauna in Garhwal Himalaya*, Sus. Eco. Sys. Env., Ed. By D.R. Khanna, A.Gautam and A.Gautam. pp: 99-104.

Badola, S.P. and Singh, H.R., (1981): Fish and fisheries of river Alaknanda. *Proc. Nat. Acad. Sci.*, 15(B): 13-142.

Badola, S.P. and Singh, H.R., (1981): Hydrobiology of the river Alaknanda of Garhwal Himalaya. *Indian J. Ecol.*, 8(2): 269-276.

Bahura, C.K., (1998): A study of physico-chemical characteristics of highly eutrophic temple tank, Bikaner. *J. Aqua. Biol.*, 13(1&2): 47-51.

Balls, H.B., and Moss, B., Irvine, (1989): The loss of submerged plants with eutrophication. Experimental design, water chemistry, aquatic plant and phytoplankton biomass in experiments carried out in ponds in the Norfolk Broadland. *Freshwater Biol.*, 22: 71-87.

Bandela, N.N., Vaidya, D.P., Shivanikar, S.V. and Lomte, V.S., (2002): Determination of Biochemical Oxygen Demand to estimate Biodegradability and self purification capacity of Burul Dam water, Nanded M.S. *J. Aqua. Biol.*, 17(1): 21-22.

Basu, M. and Mitra, A.K., (2002): Studies on the pollution of river Damodar at Burdwan with special reference to metals, COD and pH. *J. Nat Env.and Poll Tech.*, 1(4): 397-399.

Battish, D.K. and Kumari, P., (1996): Effect of physico-chemical factors on seasonal abundance of cladocera in tropical pond at village Raqba, Ludhiana. *Ecol.*, 13(1): 146-151.

Belsare, D.K., (1987): Limnology of protected water work in the Tropic. Arch. Hydrobio. Beith Ergebn. *Limnol.*, 169-178.

Belsare, D.K., Gautam, A., Prasad, D.Y. and Gupta, S.N., (1990): Limnological studies on Bhopal lakes. Seasonal changes in abiotic factors in the upper lake. *Proc. Net. Acad. Sci.*, 60: 147-150.

Berger, C., Ba, N., Gugger, M., Bouvy, M., Rusconi, F., Coute, A., Troussellier, M., Bernard, C., (2006): Seasonal dynamics and toxicity of Cylindrospermopsis raciborskii in Lake Guiers (Senegal, West Africa). *FEMS Microbiol Eco.*, 57: 355-366.

Bhatt, L.R., Lacoul, P., Lekhak, H.D. and Jha, P.K., (1999): Physico-chemical Characteristics and phytoplanktons of Taudaha lake, Kathmandu. *Poll. REs.*, 18 (4): 353-358.

Bhowmic, M.L., (1968): Environmental factors affecting fish pond in fresh water fisheries, Kalyani, W. Bengal. *Ph.D. Thesis*, Kalyani University, Kalyani, India.

Bilgrami, K.S. and Duttamunshi, J.S., (1985): *Ecology of river Ganga (Patna Farrakka)* Technical report, CSIR.

Biggs, J., Fox, G., Nicolet, P., Walker, D., Whitfield, M. and Williams, P. (1998): *A guide to the methods of the National Pond Survey*. Oxford, Pond Action.

Biggs, J., Williams, P., Whitfield, M., Nicolet, P. and Weatherby, A. (2005): *15 years of pond assessment in Britain: results and lessons learned from the work of Pond Conservation*. Aquatic Conservation-Marine and Freshwater Ecosystems 15(6): 693-714.

Birch, S. and McCaskie, J. (1999): *Shallow urban lakes: a challenge for lake management*. Hydrobiologia 395: 365- 377.

Biswas, A.K., (1979): *History of Hydrology, North – Holland*, Amsterdam.

Biswas, Asit K., (1998): *Water resources: Environmental Planning, management and development*, Tata Mc. Graw Hill Publishing Company Limited, New Delhi. pp: 1-737.

Britto., S.J., (1991): *Macrophytes collection Procedure*. St. Joseph College, Tiruchirapaily, pp. 1-33.

Brune, G.M., (1953): *Trap Efficiency of Reservoirs*. Transactions of the American Geophysical Union, 34 (3): 407-418.

Burns, D.G., Camakaris, H.M., Janseen, P.H., Dyall-Smith, M.L., (2004): Combined use of cultivation dependent methods indicates that members of most haloarchaeal groups in an Australian crystallizer pond are cultivable. *Acta Cient Venej*, 70(9): 5258-65.

Butcher, R.W., (1924): *The Plankters of the river Wharfe (Yorkshire). Naturalist*, 175-180;211-214.

Cameron, C., (1989): *Is This Any Way to Run a Reservoir?* North American Lake Management Society 9th Annual International Symposium on Lake and Reservoir Management.

Canfield, D.E., and Bachmann, R.W., (1981): Prediction of Total Phosphorus Concentrations, Chlorophyll a, and Secchi Depths in Natural and Artificial Lakes. *Canadian Journal of Fisheries and Aquatic Sciences*, 38: 414-423.

Carchini G, Solimini AG, Ruggiero A. (2005). *Habitat characteristics and odonate diversity in mountain ponds of central Italy*. Aquatic Conservation: Marine and Freshwater Ecosystems 15: 573–581.

Carlson, R.E., (1977): A Trophic State Index for Lakes. *Limnology and Oceanography*, 22: 361-369.

Carpenter, K.E., (1928): *Life in inland waters with special reference to animals*. Sidwick and Jackson, London,1-15; 1-267.

Carpenter, S.R. and Adams, M.S., (1977): The macrophyte tissue nutrient pool of a hardwater eutropic lake: implication for macrophyte harvesting. *Aquat. Bot.*, 3: 239-255.

Carpenter, S.R. and Lodge, D.M., (1986): Effects of submerged macrophytes on ecosystem process. *Aqua, Bot.*, 26: 341-370.

Chacko, P.I. and Srinivasan, R., (1955): Observations on the hydrobiology of the major rivers of Madras state, south India. *Fresh Biol. Stn. Madras,* 13: 1-14.

Chakarbarty, R.D., Ray, P. and Singh, S.B., (1959): A quantitative survey of plankters and physiological conditions of the river Jamuna at Allahabad, *1954-1955. Indian J. Fish.,* 6(1): 186-203.

Chanderashekhar, S.V.A. (1998): Rotatorian and Cladoceran fauna of Manjira Lake A.P. with a note on their abundance and indicator value. *Proc. Acad. Environ. Biol.,* 7(1): 27-30.

Chandler, D.C., (1944): Limnological studies of western lake Eris. N. Relation of Limnological and climate factors to phytoplankters of 1941. *Trans. Amer. Micros. Soc.,* 63: 203-235.

Chandler, D.S., (1940): Limnological studies of western Eric. I : Plankters and certain physico-chemical Data of Bass Island region from September 1938 to November, 1939. *Ohio J. Sci.,* 40: 291-336.

Chapman, Deborah. V., (1998): *Water Quality monitoring: in water resources,* edited by Asit K. Biswas, Tata Mc. Graw Hill Publishing House, New Delhi. pp: 209-248.

Chattopadhya, S.N., Routh, T., Sharma, V.P., Arora, H.C. and Gupta, R.K., (1984): *A short-term study on the pollution status of Ganga river in Kanpur Region.* Indian J. Environmental Hlth., 26 (3): 244-257.

Chopra, A.K. and Patrik J.N., Nirmal J. (1994):. *Effect of domestic sewage on self-putrification of Ganga water at Rishikesh.* Physico-chemical parameters. *Ad.Bids.* 13(1): 75-82.

Chopra, A.K. and Rehman, A. (1995): *A study on self-publication of Physico–chemical properties of Ganga Canal water at Jwalapur, Haridwar, Him. J. Env- Zoo,* 9: 11-13

Chugh, Tarun (2000): *Seasonal variation in the microbial ecology of River Ganga at Haridwar,* Ph. D. thesis submitted to Gurukul Kangri University Haridwar.

Cook, C.D.K. (1974): Water plants of the world. *A manual for the identification of the genera of fresh water macrophytes.* W.Junk. B.V. Publishers, The Hague, pp 561.

Cooke, G.D. and Kennedy, R.H. (2001): *Managing Drinking Water Supplies.* Journal of Lake and Reservoir Management 17 (3): 157-174.

Crayton, W.M. and Sommerfield, M.K. (1979): *Composition and abundances of phytoplankters in tributaries of the lower Colorado River, Canyon region,* Hydrobiologia, 66(1): 81-93.

Crottereau, C, Delmas, D. (1999): *Regulation of bacterial proteolytic activity at natural concentrations in dissolved organic matter in a maritime pond.* C.R. Acad Sci III, 322 (7): 551-6.

Daborn, G.R. (1976): *Physical and chemical factors of a vernal temporary pond in western Ganaga:* Hydrobiologia, 51 (1): 33-38.

Das, A.K., (2000): *Limno-chemistry of some Andhra Pradesh reservoirs.* J. Inland. Fish Soc. India, 32: 37- 44.

Das, H.B., Kalita, H., Saikia, L.B., Borah, K. and Goswami, C.C. (1989): *Portability of waters of Arunachal Pradesh.* JIWWA, 21: 191.

Das, S.M. (1989): In: *Handbook of Limnology and water pollution.* South Asian Publishers, New Delhi 1-174

Das, S.M. and Akhtar, S. (1970): *A report on fresh water plankters from Dul Lake*, Kashmir. Sc., 7 (1-2): 133-137.

Das, S.M. and Shrivastava, V.K. (1959): *Studies on freshwater plankton III. Qualitative, composition and seasonal fluctuation in plankton components*, Proc. Nat. Acad. Sci. India, 29: 174-189.

David, A., Rao, N.G.S. and Rahman, F.A., (1969): *Limnology and fisheries of Tungabhadra reservoir.* Bull C.I.F.R.I. Barrachpore, 13:188.

Davies, B., Biggs, J., Williams, P., Whitfield, M., Nicolet, P., Sear, D., Bray, S. and Maund, S. (2008). *Comparative biodiversity of aquatic habitats in the European agricultural landscape.* Agriculture Ecosystems & Environment 125(1-4): 1-8.

Davis, G.C. (1955): *The marine and fresh water plankton.* Michigan State University Press, East Lansing.

Dawidowicz, P and Gliwicz, Z.M, (1987): Biomanipulation-III, 33: 259-278.

De, Jong, L, Moreau, X, Jeans, S., Thiery, A. (2006): *Expression of the heat shock protein Hsp 70 in chloride target cells of mayfly larvae from motorway retention pond; a biomarker of osmotic shock.* Sci.8 Total Environ; 31:366 (1):164-73.

De Meester L, Declerck S, Stoks R, Louette G, Van de Meutter F, De Bie T, Michels E, Brendonck L. 2005. *Ponds and pools as model systems in conservation biology, ecology and evolutionary biology.* Aquatic Conservation: Marine and Freshwater Ecosystems 15: 715–725.

Declerck, S., and others. 2005. *Multi-group biodiversity in shallow lakes along gradients of phosphorus and water plant cover.* Ecology 86:1905-1915.

Deevey, G.B. (1948): *The Zooplankters of Tisbury Great pond. Bull.* Bingham Oceanogr. Coll., 12 (1): 1-44.

Della Bella V, Bazzanti M, Chiarotti F. (2005). *Macroinvertebrate diversity and conservation status of Mediterranean ponds in Italy: water permanence and mesohabitat influence.* Aquatic Conservation: Marine and Freshwater Ecosystems 15: 583–600.

Derome, N., Duchesne, P., Bernatchez, L., (2006): *Parallelism in gene transcription among sympatric lake whitefish (Coregonus clupeafoirmis Mitchill) ecotypes.* Mol. Ecol., 15 (5): 1239-49.

Desikachary, T.V. (1959): *Cyanophyta Indian council of agricultural research.* Monographs on Algae No.2

Dobriyal, A.K. and Singh, H.R. (1981): *Diurnal variation in some aspects of limnology of the river Mandakini from the Garhwal Himalaya, Uttar Pradesh* J.Zool., 1: 16-18.

Dobriyal, A.K., Singh, H.R. and Bisht, K.L. (1983): *Diurnal variation in hydrobiological parameters of two hill streams of Garhwal Himalaya. Uttar Pradesh* J. Zoo., 3:30-34.

Dorris, T.C. (1963): *Limnology of the middle Mississippi river IV physical and chemical limnology of river and chute.*, Limnol. Oceanogr, 8:79 – 99.

Doudoroff, P. (1957): *The physiology of fishes* Vol. (2) Academic Press, New York.

Douglas, M.S., Smol, J.P., Savelle, J.M., Blais, J.M. (2004): *Prehistoric Inuit whalers affected Arctic freshwater ecosystem.* Proc. Nat Acad. Sci. U.S.A.:101 (6):1613-7.

Driuban, A.N. (2000): *Bacterioplankton and bacteriobenthos of some floodplain lakes in the course of lower Amur River.* Mikrobiologia, 71 (4): 550-7.

Dube, P., Sharma, Y., Sharma, K.K. (Dept Zoo, MDS Univ, Ajmer 305009 Rajasthan), (2002): *Ecobiology of seasonal waterbodies in southeastern plateau of Rajasthan with special reference to amphibians.* Indian J Environ Sci, 6(2): 135-139.

Dubrovin, T., Jolma, A. and Turunen, E., (2002): *Fuzzy model for real time reservoir operation.* Water Resources Planning and Management 128(1): 66-73.

Dutta, N.C., Bandopadhaya, B.K. and Bandopadhyaya, S.B. (1983a): *Relationship between the phytoplankters and primary productivity of a freshwater pond at Calcutta.* Env. Eco., (3):137

Dutta, S.P.S., Malhotra, Y.R., Sharma, K.K., and Sinha, K. (1997): *Diel variations in physico-chemical characteristics of water and macrobenthic invertebrates in river Tawi, near Nagrota by pass, Jammu.* Him. *J. Env. Zool.*, Vol.11, pp.45-50.

Dwivedi, B.K., Pandey, G.C. (Dept Environ Sci, Dr RML Avadh Univ, Faizabad) (2002): *Physico-chemical factors and algal diversity of two ponds (Girija Kund and Maqubara Pond), Faizabad, India.* Polln Res, 21(3): 361-370.

Dwivedi, B.K., Pandey, G.C. (2003): *Complex dynamics of toxin producing algal species and primary productivity in two water ponds of FAizabad.* J. Environ. Biol, 24(1):55-61.

Eddy, S. (1934): *A Study of the fresh water Plankters Communities Illinois*, Biol. Monogr, 12 (4): 1-93.

Edmondson, W.T. (1961): *Changes in lake Washington following an increase in nutrient income;* Verth. Int. Ver. Limnol., 14: 167-175

Edmondson, W.T. (1977): *Population dynamics and secondary production:* Ergeb. Limnoi., 56-64.

Egborge, A.B.M. (1978c): *Observations on the diurnal changes in some physicochemical variables of Lake Asejire, a new impoundment in Nigeria.* Pol. Arch. Hydrobiol., 26(2):61-73.

Egborge, A.B.M. (1979): *The effect of impoundment on the water chemistry of Lake Asejire, Nigeria.* Freshwater Biol. 9.403-412.

Egborge, A.B.M. (1986): *Physical, chemical and biological properties of Benin River. Technical Report* for Cage Fishery Development at Kolo Submitted to Nr. J. Ayomike, p. 32.

Eggleston, J.R., (2003): *Evaluation of Strategies for Balancing Water Use and Streamflow Reductions in the Upper Charles River Basin, Eastern Massachusetts. Water-Resources Investigations Report* 03-4330, 85 pp.

Esmaeili, H.R. and Johal, M.S. (2005): *Study of physico-chemical parameters of water of Govind Sagar Reservoir, India,* Proceedings of National Seminar "New trends in fresh development in India". Punjab University pp. 173-177.

Ernst, W., Doe, K., Jonah, P., Young, J., Julien, G., Hennigar, P. (1991): *The toxicity of chlorothalonil to aquatic fauna and the impact of its operational use on a pond ecosystem.* Arch. Environ. Contam. Toxicol. 21(1):1-9.

Fasset, (19750): *A manual of aquatic vascular plants.* The University of Wisconsin press, Madison. pp. 405.

Fokmare, Anil, K. and Musaddiq, Mohmmad. (2002): *A study on Physico-chemical characteristics of Kapshi lake and Purna river water in Akola district of Maharashtra (India).,* J. Nat. Env. and Poli.Tech., 1(3): 261-263.

Forel. F. A. (1892): *Le Leman. Monographic* Limnogique, 3 Vols. Lavsane.

Foy, R.H. (1986): *Suppression of Phosphorus Release from Lake Sediments by the Addition of Nitrate.* Water Research 20 (11): 1345-1351.

Franklin, T. (1969): *Phytoplankters of Bhavani Sagar reservoir, Madras.* J. Fish.,5:112-118.

Fulmer, D.G. and G.D. Cooke. (1990): *Evaluating the Restoration Potential of 19 Ohio Reservoirs.* Lake and Reservoir Management 6 (2): 197-206.

Gambhir, Sanjay Kumar (1999): *Physico-chemical and biological characteristics of water of Maithon reservoir of D.V.C.* Poll. REs. 18 (4):541-544.

Ganapati, S.V. (1943): *A ecological study of a garden pond containing abundant zooplankters.* Proc. Ind. Acad. Sci., 17: 41-58.

Ganapati, S.V. (1964): *A five year investigation of Almati reservoir.* Part IV, Zooplankters. Env. Hlth., 6: 65-74.

Ganapati, S.V. (1966): *The rotorianfauna of Ajwa reservoir and the Nimeta water works at Baroda.* Limnologia Berlin, 4:55-93.

Gautam A. (1988): *Limnological studies and seasonal variations in the plankters organisms and primary production in the upper lake of Bhopal.* Ph.D. thesis Bhopal University, Bhopal 140 pp.

Gautam, A. (1990): *Ecology and pollution of mountain waters,* Ashish Pub. house. N. Delhi. pp.206

George, J.P., (1976): *Hydrobiological studies on the lower lake of Bhopal with special reference to the productivity of economic fishes.* Ph.D. thesis Bhopal University, Bhopal.

George, M.G. (1966): *Comparative plankters ecology of five tanks in Delhi.* Hydrobiologia. 27(2):81-108.

Gledhill, D. G. and James, P. (2008). *Rethinking Urban Blue Spaces from a Landscape Perspective: Species, scale and the human element.* Salzburger Geographische Arbeiten 42: 151-164.

Goel, P.K., and Trivedi, R.K. (1984): *Chemical and biological methods for water pollution studies.* Environmental Publications Karad 250pp.

Goldman, C.R. and Horne, A.J. (1983): *Limnology international student edition,* McGrew Hill International Book Company, Tokyo, Japan.

Goldman, C.R. and Wetzel, R.G. (1963): *A study of primary productivity of clear Lake,* Lake County, California. Ecology, 44: 283-294.

Golterman, H.L. (1976): *Influence of soil on the chemistry of water in relation to productivity.* F.A.O. Fish Rep., (44) 3: 27-42.

Gonzalez, F., Zoppi, de., Roa, E. (1999): *Chemical disturbance of the phytoplankton structure in an artificial pond.* Acta. Cient. Venez. 50 (4): 195-200.

Gopal, B. Goel, P.K., Sharma, K.P. and Trivedy, R.L. (1981): *Limnological study of fresh water reservoir, Jamwa Ramgarh (Jaipur), Hydrobiologia,* 83: 283-294.

Gottschalk, L.C., (1952): *Measurement of sedimentation in small reservoirs, Trans Am Soc. Civil Engg.,* 117: 59-71.

Govind, B.V. (1963): *Preliminary studies on plankters of the Tungabhadra reservoir.* Indian J. Fish. 10(1): 148-158.

Green, J. (1960): *Zooplankters of river Skoto. The Rotifera.* Proc. Zool. Soc. London.125: 491-523.

Gupta, S. (1989): *Pollution ecology of some ponds in urban vicinity of Jodhpur* Ph.D. Thesis University of Jodhpur 1-234.

Harvey, H.W., Copper, L.H.N., Lebour, M.V. and Russel, F.S. (1935): *Plankters production and its control.* J.Mar. Biol. Assoc. U.K., 20: 407-462.

Hasan, G.O., Paul P. Mathisen and Don Pellegrino (2007): *Distribution of heavy metals in vegetation surrounding the Blackstone River, USA: Considerations regarding sediment contamination and long term metals transport in freshwater riverine ecosystems.* J. Environ. Biol., 28, 493-502.

Haslam, S.M. (1978): *River plants. The macrophytic vegetation and their identification and management,* Cambridge University Press, Cambridge.

Herzig, A. (1979): *Neusiedlersee: in the limnology of a shallow lake in the central Europe.* In Ed. H.Loffler (Ed.):282-335.

Hinden H, Oertli B, Menetrey N, Sager L, Lachavanne JB. (2005). *Alpine pond biodiversity: what are the related environmental variables?* Aquatic Conservation: Marine and Freshwater Ecosystems 15: 613–624.

Hussaini, S.U. (1967): *Studies on the limnological and primary production of tropical lake.* Hydrobiologia. 30 (3-4): 335-352.

Hutchinson, G.B. (1957): *A Treatise on Limnology Vol. 1: Geography, Physics, and Chemistry.* John Wiley and Sons, Inc., New York: 115.

Hutchinson, G.E. (1967): *A treatise on limnology. Vol II. Introduction to Lake biology and the Limnoplankters.* John Wiley and Sons, Inc. New York, 1115 pp.

Hynes, H.B. N (1970): *The Ecology of Running Water.* Liverpool University Press, Liverpool. Pp 555.

Hynes, H.B.N. (1960): *The Biology of Polluted Waters.* Liverpool University Press, Liverpool, pp. 431.

Imevbore AMA (1964): *Studies of the zooplankters and physico-chemical changes in a tropical freshwater habitat at Ibadan, Nigeria.* Ph.D. thesis, University of London.

Jackson, D.P. (1961): *Comparative study on phytoplankters photosynthesis in relation to total alkalinity:* Ver. Int. Ver. Limnol., 14: 125-135.

Jeejabai, N. and Rajendra, S. (1982): *Phytoplankters composition as indicators of water quality- A study of Adyar river. Proc. Muragappa Chettiar Symposium.*

Jena, B., Sudarshan, R. and Chaudhary, S.B. (2003): *Studies on water quality parameters around Sagar island, Sunderbans, J. Nat env. and Poll. Tech.*, 2(3): 329-332.

Jenkin, P.M. (1942): *Seasonal Changes in the temperature of Windermere (English Lake District).* J. Anim. Ecol., 11: 248-269.

Jewell, W.J. (1971): *Aquatic weed decay: dissolved oxygen utilization and nitrogen and phosphorous regeneration.* J. Water Pollut. Control Fed. 43, 1457-1467.

Jhajharia, Anita, (2002): Limnobiotic studies of Sekhawat Pond, Jodhpur with special reference to plankton. *J. Nat. Con.*, 14(2): 215-224.

Jhingran, V.G. (1975): *Fish and Fisheries of India,* Hindustan Publishing Corporation, (India): 1-954.

Jhingran, V.G. (1982): *Fish and fisheries of India 2nd edition.* Hindustan Publishing Corporation (India), Delhi.

Joshi, B.D., Bisht, R.C.S. and Joshi Namita, (1996): *Plankters population in relation to certain physico-chemical factors of Ganga canal at Jwalapur (Haridwar).* Him. J. Env. Zool. Vol. 10: 75-77.

Joshi, B.D., Pathak, J.K., Singh, Y.N., Bisht, R.C.S. and Joshi, P.C. (1993): *On the physico-chemcial characterstics of river Bhagirathi in uplands of Garhwal Himalaya.* Him. J. Env. Zool., 7:64-75.

Joshi, G. (1987): *Diurnal studies of Physico-chemical Properties of Limnological importance.* Ph.D. Thesis, Dr. Harsingh Gaur Vishwavidyalaya, Sagar, 204 pp.

Kala, Rakhee and Sharma, R.C. (2002): *Effect of Physico-chemical factors on phytoplankters population in lotic environment of Alaknanda river, Garhwal Himalaya,* Indian *J. Ecol.*, 29(2): 221-226.

Kalyani, Y. and M.A. Singara Charya (1999): *Phytoplankton dynamics in Bhadrakali lake, Warangal (A.P.) In: Fresh water ecosystem of India.*(Ed. K. Vijay Kumar) Daya Publishing House, New Delhi. Pp. 209-225.

Kannan, V. and Job, S.V. (1980): *Diurnal depthwise and seasonal changes of physico-chemical factors in Sathiar reservoir.* Hydrobiologia. 70: 103-117.

Kant, S. and Anand, V.K. (1978): *Interrelationship of phytoplankters and physical factors in Mansar lake, Jammu, (J&K). Indian J. Ecol.*, 5(2): 134-140.

Kaul, S. (1983): *Limnological studies of an aquatic body. Lalpari Talab at Rajkot,* Indian J. Ecol., 10 (1): 124-128.

Kaur, H., Dhillon, S.S., Kaur, N. and Bath, K.S. (1995): *Physico-chemical charaterstics of freshwater ecosystem prevailing in and around Patiala (Punjab),* J. Environ. and Ecology. 13 (3): 523-528.

Kaur, Rajendra and Khulbe, R.D. (1998): *Occurrence, Distribution and ecology of watermolds in Nanaksagar reservoir Kumaun Himalaya Nainital,* Uttar Pradesh India, J. Indian bot. Soc., 17: 111-116.

Kaushik, S. and Saksena, D.N. (1999): *Physico-chemical limnology of certain water bodies of central India.* In: *Fresh water ecosystems of India* (Ed. K. Vijay Kumar) Daya Publishing House, New Delhi, pp. 158.

Keating, K.I. (1978): *Blue Green Algae inhibition of diatom growth transition from mesotrophic to eutrophic community structure.* Science, 199: 991-973.

Khan, A.A. and Siddiqui, A.D. (1970): *Diurnal Variation in the pond moat at Aligarh.* J. Inl. Fish Soc. India. 11: 146-154.

Khan, A.A. and Siddiqui, A. (1974): *Seasonal changes in the limnology of a perrennial fish pond at Aligarh.* Indian J, Fish, 21(2): 463-478.

Khan, A.A., Siddiqui, A. (1971): *Primary Production in a tropical fish pond at Aligarh India.* Hydrobiologia. 37 (3-4): 447-456.

Khan, M.A. and Zutshi, D.P. (1980): *Primary Productivity and trophic status of a Kashnir Himalaya lake.* Hydrobiologia., 68 (1): 3-8.

Khanna, D.R., (1993): *Ecology and pollution of Ganga River.* Ashish Publication House, Delhi. pp: 1-241.

Khanna, D.R., Badola, S.P. and Dobriyal, A.K. (1993): *Plankters Ecology of the River Ganga at Chandighat, Haridwar,* Advances in limnology, Ed. By H.R. Singh, Narendra Publishing house, New Delhi. 171-174.

Khanna, D. R. and Chugh, T. (2004): *Microbial Ecology: A study of River Ganga,* Discovery publishing house, New Delhi. 1-277.

Khanna, D. R., Badola, S.P. Singh, H.R. and A.K. Dobriyal (1992): *Observations on seasonal trends in diatomic diversity in the river Ganga at Sapt-sarover, Haridwar.* Recent researches in cold water Fisheries: 99-107.

Khanna, D.R. and Badola, S.P. (1990): *Ichthyofauna of the river Ganga at foothills of Garhwal Himalayas.* Journal of Natural and Phy. Sci., Vol. 4(1-2) pp153-162.

Khanna, D.R. and Bhutiani, R. (2003): *Ecological Status of Sitapur Pond at Haridwar (U.A.) Inida.* Indian J. Environ. and Ecoplan. 7(1):175-178.

Khanna, D.R. and Bhutiani, R. (2003): *Limnological characteristic of the river Ganga at Haridwar (Uttaranchal),* U.P.J.Zool., Vol. 23(3): 179-183.

Khanna, D.R. and Bhutiani, R. (2003): *Limnological status of Satikund pond at Haridwar* (U.A.), *Indian J. Env. Sc.*, Vol.7 (2): 131-136.

Khanna, D.R. and Bhutiani, R. (2004): *Fish and their ecology of River Ganga at Gohri Ghat Garhwal (Uttaranchal), Proc. Nat. Sem, Fish diversity in protected habitats*, Ed. By S. Ayyappan, D.S. Malik, Rani Dhanze and R.S. Chauhan pp. 291-298.

Khanna, D.R. and Bhutiani, R. (2006): *Water analysis- At a Glance.* ASEA publication, pp.1-116.

Khanna, D.R. and Bhutiani, R. (2007): *Limnological Modeling: A case Study of River Suswa.* Daya Publishing House: 1-297.

Khanna, D.R. and Singh R.K. (2000): *Seasonal fluctuations in the plankters of Suswa river at Raiwala (Dehradun).*, Env. Cons. J., Vol. 1 (2and3): 89-92.

Khanna, D.R. and Singh, S. (2002): *Pond fish Ecology and Economics*, Daya Publishing House, Delhi, 1-816.

Khanna, D.R. and Singh, Shekhraj. (1999): *Diurnal variation in microbial ecology of Narkewala Pond at Ferupur, Haridwar*, Sus. Eco. Sys. and Env., Ed. By D.R. Khanna, and A. Gautam, pp. 99-104.

Khanna, D.R., Chugh, Tarun and Sarkar Praveen (2001): *Fluctuation in the population density of Macro-invertebrates of river Ganga At Pashulok barrage, Rishikesh (Uttaranchal, India), Env. Cons. J.*, 2(1): 37-39.

Khanna, D.R., Malik, D.S. and Rana, D.S. (1997): *Phytoplankters communities in relation to certain physico-chemical parameter of Ganga canal at Haridwar., Him. J.Env- Zool.*, 12; 193-197.

Khanna, D.R., Gautam, A., Chugh, T., and Sarkar, P. (2000): *Impact of abiotic factors on the phytoplankton population of a pond.* Environmental Conservation Journal. (1): 41-46.

Khanna, D.R., Gautam, Ashutosh, Chugh, Tarun, and Sarkar, Praveen (2000): *Impact of abiotic factors on the phytoplankters population of a pond*, Env. Con. J., 1(1): 41-46.

Khanna, D.R., Malik, D.S. and Badola, S.P. (1997): Population of green algae in relation to physico-chemcial factors of the river Ganga at Lal-Ji-Wala, Haridwar. Uttar Pradesh J. Zool. 17(3): 237-240.

Khanna, D.R., Malik, D.S., Seth T.R. and Rupendra (1999): Correlation between abiotic factors and plankters population in river Ganga at Rishikesh (U.P.). Sus. Eco. Sys. and Env. Ed. By D.R. Khanna, A. Gautam and A. Gautam, Published by ASEA, Rishikesh. P. 69-76.

Khanna, D.R., Malik, D.S., Seth, T.R. and Rupendra (1999): *Correlation Between abiotic factors and plankters population in river Ganga at Rishikesh (U.P)*, Sus. Eco Sys. and Env, Ed. by D.R.Khanna, A Gautam and A. Gautam, Published by ASEA, Rishikesh pp. 69-76.

Khanna, D.R., Seth, T.R., Malik, D.S. and Singh, R.K. (1999): *Fishes and their ecology of river Suswa at Raiwala (Dehradun)*, Sus. Eco. Sys. and Env., Ed. By D.R. Khanna, A.Gautam and A.Gautam pp-105-110.

Khare, G.K. and Sastri, C.A. (1970): Studies in characterization and pollution effects of viscope rayon waste. Environ. Hlth., 12 (2):11-19.

Khatri, T.C. (1980): Limnological studies of Lakotia Lake Pali, Rajasthan Ph.D. Thesis, Univ. of Jodhpur, India 134.

Khulbe, R.D. and Durgapal, A. (1993): Evaluation of drinking water quality at Bhimtal, Nainital, Uttar Pradesh. Poll. Res. 21 (2): 109-111.

Kietlinska, A., Renman, G., Janness, Tham, G. (2005): *Nitrogen removal from landfill leachate using a compact constructed wetland and the effect of chemical pretreatment.* J. Environmental Sci. 40(6-7):1493-506.

Klein, L. (1957): *Aspects of river pollution.* Butterworth Scientific Publication.

Kohli, M.P. (1981): *Plankters Study of Gobindsagar reservoir (India).* Comp. Physiol. Ecol., 6(1):49-52.

Kudesia, V.P. and Verma, S.P. (1996): *Physico-chemical studies in industrial pollution of Kalinadi due to combined effluents of canesugar, chemical industry, distillery and rubber industries in Meerut region.* Indian. J. Env. Agric. 1(1): 1-11.

Kulsheshtra S.K. (1988): *Review on studies of Bhopal Lakes.* Proc. Nat. Symp. Past. Present and Future of Bhopal Lakes. 11-22.

Kumar, Arvind (1995): *Studies in pollution in river Mayurakshi in South Bihar.* Indian J. Env. Poll. 2(1) 21-26.

Kumar, Neeraj and Sharma, R.C. (2002): *Water quality of river Krishni {Part-I Physico-chemical characteristics}* J.Nat.Con., 14(2): 273-297.

Kumar, Neeraj and Sharma, R.C. (2002): *Water quality of river Krishni {Part-II Physico-chemical characteristics,*J.Nat.Con., 14(2): 299-333.

Lakshminarayana, J.S. (1975): *Prevention of sewage pollution by stabilization of ponds.* Environ Lett. 8(2): 121-34.

Lewis, W.M. (1974): *Primary production in the plankters community of a trophical lake.* Ecol. Monogr. 44 (4): 337-409.

Lewis, W.M. (1979): *Zooplankters Community Analysis.* Springer Verlag, New York 163 pp.

Lind, O.T. (1979): *Handbook of common methods in Limnology.* The C.V. Mosby Co.2nd. edition. St. Louis, Missouri 199 pp.

Lindman, R.L. (1942): *The Trophic dynamic aspect of ecology.* Ecology. 23:399-418.

Lund, J.W.B. (1965): *The ecology of freshwater phytoplankters.* Biol. Rev., 40: 231-293.

Lund, J.W.G. (1964): *Primary Production and periodicity of phytoplankton.* Verb. Int. Ver. Theor. Angew. Limnol., 15: 37-56 (M.P.). In "In Current Trends in Limnology I" (Nalin Kshastree ed.) pp. 159-184, Narendra Publishing House, New Delhi,

Lund, J.W.G. (1972): *Eutrophication.* Poc. R. Soc. London Ser. B.,180: 371-382.

Macan, T.T. (1974): *Fresh water Ecology,* 2nd Edition, New York: Wiley, 224pp.

Mackereth FJH (1963): *Some methods of water analysis for limnologists.* Freshwater Biol. Ass, Sci. Pub. 21:72.

Mahajan, A. and Kanher, R.R., (1995): *Variation in primary productivity of a fresh water pound in west Nimar (M.P.).* Acta Ecol.,17(2): 148-156.

Menetrey N, Sager L, Oertli B, Lachavanne JB. (2005). *Looking for metrics to assess the trophic state of ponds. Macroinvertebrates and amphibians.* Aquatic Conservation: Marine and Freshwater Ecosystems 15: 653–664.

Margalef (1969): *Diversity and Stability in ecological system.* BNO publication 22.

Maruthanayagam, C., Sasikumar, M. and Senthilkumar, C. (2003): *Studies on Zooplankters population in Thirukkulam pond during summer and rainy seasons.,* J. Nat. Env. and Poll. Tech., 2(1): 13-19.

Mathews, W.J. (1998): *Patterns in Freshwater Ecology.* Chapman and Hall, International Thomas Publishing.

Mc Comble, A.M. (1953): *Factors influencing the growth of phytoplankters.* J. Fish Res. Bd. Canada, 10 (5): 253-282.

Meysman, F. J. R., Galaktionov, O. S., Gribsholt, B., and Middelburg, J. J., (2006): *Bio-irrigation in Permeable Sediments: Advective Pore Water Transport induced by Burrow Ventilation.* Limnol. Oceanogr., 51: 142–156.

Meysman, F.J.R., Galaktionov, O.S., Cook, P.L.M., Janssen, F., Huettel, M. and Middelburg, J.J., (2007): *Quantifying biologically and physically induced flow and tracer dynamics in permeable sediments.* Biogeosciences, 4: 627-646.

Michael, R.G. (1968): Studies on bottom fauna in tropical fresh water pond. Hydrobiologia. 31: 203-230.

Michael, R.J. (1964): *Diurnal Variation of plankters correlated with physico-chemcial factors in three different ponds.* Ph.D. Thesis, University of Calcutta, Calcutta.

Michael, R.J. (1968): *Studies on Zooplankters of a fish pond.* Hydrobiologia, 32: 47-68.

Mishra, A.P., B.K. Borah and Sharma, M. (1999): *Limnological investigations of a fresh water tributary.* J. fresh water Biol. 11(1-2): 1-5.

Mohanty, R.C. (1981): *Water quality studies of some water bodies of Bhubnaeshwar,* Ph.D Thesis, Utkal University, pp. 1-240.

Moitra, S.K. and Bhowmik, M.L. (1968): *Seasonal cycles of rotifers in a fresh water fish pond in Kalyan, West Bengal.* Proc. Symp. Adv. Trop. Ecol., Part I: 359-367.

Monichan, K.K. and Sharma, R.K. (2003): *Study on the distribution status of fish fauna in Etawah district, U.P.,* Indian J. Env. Sc., 7(2): 115-117.

Monizsette, D.G. and Mavinic, D.S. (1978): *BOD Test Variables.* J. Envir. Eng. Div., EE6: 1213-1232.

Moore, K., Jones, K. and Dambacher, J. (1995): *Methods of stream habitat surveys,* version 5.1 Oregon Dept. of Fish and Wildlife, Corvallis.

Mortimer, C.H. (1942): *The exchange of dissolved substance between mud and water in lakes* part II. J. Ecol., 30: 147-201.

Mruthunjaya, T.B. and Hosmani, S.P. (2002): *Comparative account of physico-chemical parameters, enzymatic activity and plankters population in three water bodies of Mysore city.*, J. Nat. Env. and Poll. Tech., 1(3): 165-169.

Mujumdar, V. (1979): *Physico-chemical aspects of Bhad Bhada Fish Farm M.Phil.* Thesis. Bhopal Univ. M.P. 1-43.

Muller, G. and Forstner, U (1968 b): *General relationship between suspended sediment concentration and water discharge in Alpien Rhein and some other Rivers.* Nature 217: 244-245.

Munawar, M. (1970): *Limnological Studies on fresh water ponds of Hyderabad, India, I.* The Biotop. Hydrobiologia, 35 (1): 127-162.

Munshi, J.L. and Munshi, J.S.D. (1995): *Fundamentals of fresh water Biology.* Narendra Pub. House, Delhi, India.

Murugesan, A.G., Abdul Hamed, K.M.S.A. and Sukumaran, N. (1994): *Water quality profile of the perennial river Tampraparni.* Indian J. Env. Prot., Vol. 14(8): 567-572.

Nalewazko C., Bryant, G. and Sreenivasan, M. (1981): *Limnology of Heart Lake.* Ontario. Hydrobiol., 79: 245-253.

Narayana, J and R.K. Somashekar, (2002): *Macrophytes diversity in relation to water quality – Investigation on river Cauvery.* In: Ecology and conservation of lakes, reservoirs and rivers. ABD Publishers, Jaipur.

Narayana, J. and R.K. Somashekar, (1997): *Heavy metal composition in the sediment and plants of the river Cauvery.* Journal of Environment and Pollution, 4(4), 325-328.

Nautiyal, P. (1984): *Studies on the rivrine ecology of torrential waters in the Indian uplands of the Garhwal region II:* Seasonal fluctuations in diatom density. Proc. Indian Acad. Sci., 93 (7): 671-674.

Nayar, C.K.G. (1965): *Cyclomorphosis of Brachionus Calyciflorus Pallas.* Hydrobiologia, 25 (3-4):538-544.

Naz, Sabrina, Shah, M. Alam, and Das Tapoti (2002): *Aquatic fungi as indicator organisms.*, J.Nat. Env. and Poll. Tech., 1(3): 207-211

Needam, J.G. and Needham, P.R. (1972): *A guide to the study of freshwater biology,* Holden-Day I.N.C. san Francsco, *Calf.*, 94(3): 1-108.

Negi, K.S. and Malik, D.S. (2006): *Piscine diversity of Hillstream Chhirapani in Champawat District of Kumaon Himalaya,* J. Mountain Res., 1: 129-133.

Nygard, G. (1949): *Hydrobiological studies in Danish ponds and lakes. II The quotient hypothesis and some new or little known phytoplankters organisms.* Det Kongelige. Danske videns akbernes Selskad;Bio.; Skrifter, 7:1-293.

Odum, E.P. (1971): *Fundamental of ecology 3rd edn Saunders.* Philadalphia. 574 pp.

Oertli B, Auderset Joye D, Castella E, Juge R, Lehmann A, Lachavanne J-B. 2005. PLOCH: a standardized method for sampling and assessing the biodiversity in ponds. Aquatic Conservation: Marine and Freshwater Ecosystems 15:665–679.

Oertli B, Auderset Joye D, Castella E, Juge R, Lachavanne J-B. (2000). *Diversite´ biologique et typologie e´ cologique dese´ tangs et petits lacs de Suisse.* Swiss Agency for the Environment, Forests and Landscape, Laboratory of Ecology and Aquatic Biology, University of Geneva.

Oertli, B., Cereghino, R., Hull, A. and Miracle, R. (2009). *Pond conservation: from science to practice.* Hydrobiologia 634(1): 1-9.

Pailwan, I.F. (2005): *Limnology and fisheries potential of perennial tanks of Kolhapur district, Kolhapur,* Ph.D. Thesis, Shivaji University. Kolhapur.

Palmer, C.M. (1969): *A Composite rating of algae tolerating organic pollution.* J. Phycol. 5: 78-82.

Pandey, A.K., Siddique, S.Z. and Ramarao, K.V. (1993): *Physico-chemcial and biological characterstics of Hussain Sagar, an industrial polluted Lake, Hyderabad.*

Park, Bae Kyung and Seok Soon Park (2005): *Effects of stream hydraulic conditions on foraging strategies of false dace, Pseudorasbora parva in the lentic ecosystem.* J. Environ. Biol., 26, 635-643.

Parkash (1982): *Water quality of Kittham lake (Soor Sarovar)* J. Env. Biol. 4(4): 193-200.

Pathal, S.K. and Mudgal, L.K. (2005): *Limnology and biodiversity of fish fauna in Virla Reservoir, M.P. India,* Env. Con. Jour. Vol. 6 (1): 41-45 pp.

Patki,, Saroja S. (2002): *Hydrobiological studies of Banshelki dam at Udgir, Maharashtra.* Ph.D. thesis submitted to Dr. Babasaheb Ambedkar Marathwad University, Aurangabad.

Patrik, R. (1948): *Facotrs effecting the distribution of diatom.* Bott. Rev. 14 (8).

Patrik, R. (1955): *Diatoms as an indicator of river changes.* Proc. 9th industrial waste confpurdue, Univ. Engg., Buli., 39 (2):325.

Pawar, B.A. and U.H. Mane (2006): *Hydrography of a Sadatpur lake near Pravaranagar, Ahemednagar District, Maharashtra.* J. Aqua. Biol. Vol. 21 (1): 101-104.

Pearsall, W.H. (1930): *A theory on diatom periodicity.* Proc. Rev. Soc., 11:105-183.

Pennak, R.W. (1949): *Annual Limnological Cycles in some Colorado reservoir and lakes.* Ecol. Monogr., 91:233-267.

Pennak, R.W. (1955): *Comparative limnology of eight Colorado mountain Lakes.* Univ. Colo Stud. Biol. 2: 1-75.

Pennak, R.W. (1978): *Freshwater invertebrates of United States.* Ronald Press. Co. New York. 769 pp.

Pereira, A., Tassin, B., and Jorgensen, S.E, (1994). *A model for decomposition is the down vegetation in an Amazonian reservoir* Ecol. Modeling – 75176, 447-458.

Peterson, C.G.J. (1922): *On the stock place fisheries in different water.* Rep. Danish Biol. Sta., 29:1-43.

Phillips, C.P. C. (1927): *Diurnal Fluctuation in the hydrogen ion activity of a Minnesota Lake.* Ecology, 8:73-89.

Piska, R.S. (2000): *Concepts of Aquaculture.* Lahari Publications, Hyderabad.

Pokhriyal, R.C., Dobriyal, A.K. and Singh, H.R. (1983): *Seasonal variation in potamological parameter of the steam Khandagad. Uttar Pradesh* J. Zool., 3: 70-72.

Powers, E.R. (1938): *Factors involved in the mortality of fishes. Trans. Am. Fish. Soc.* 67:271-280.

Prasad, Bijay Bhushan, and Singh, R.B. (2003): *Composition, abundance and distribution of phytoplankters and zoobenthos in a tropical water body.,* J. Nat Env. and Poll. Tech., 2(3): 255-258.

Prasad, R.P. (1956): *Further studies on the plankters of the inshore waters of Mandapam.* Indian J. Fish., 3:1-42.

Prati, L., Pavanello, R. and Pesarin, F. (1977): *Assessment of surface water quality by a single index of pollution.* Water Res., 5: 741-751.

Prescott, G.W. (1939): *Some relationship of phytoplankters to limnology and aquatic biology.* A.M. Assoc. Adv. Sci. Publ., 10: 65-78.

Prescott, G.W.A. (1951): *Algae of the Western Great lakes area.* Institute of Science, Bulletin, No. 31: 346 pp.

Raina, V., Saha, A.R. and Ahmed, S.R. (1984): *Pollution Studies on river Jhelum – I: An assessment at water quality,*Indian J. env. Hlth., 26: 187-210.

Rao, D.S. and Govind, B.V. (1964): *Hydrobiology of Tungbhadra reservoir,* Indian J. Fish. 2(1): 321-344.

Rao, K.S. and Shrivastava, S. (1989): *Studies on biological monitoring of water quality in Chambal and Khan rivers of Central India,* Geobios, 16: 78-82.

Rao, K.S., KArtha, K.N., Gupta, D.R., Pandya, S.S. and Iyer, H.K. (1988): *Studies on morphometry and hydrobiology of Gandhi Sagar Reservoir with special reference to its fisheries,* J. Fish Tech., 25(1): 21-28.

Rao, N.G. (1987): *Synecology of the lake Rangasagar in relation to the limnology and eutrophication.* Ph.D. Thesis Sukhadia Univ. Udaipur.

Rawson, D.A. (1960): *A limnological comparison of twelve large lakes in northern Saskatchewan.* Limnoloceanogr, 5:195-211.

Rawson, D.S. (1951): *The total mineral content of lake water.* Ecology, 32: 669-672.

Ray, H.K. (1955): *Plankters ecology of the river Hoogly at Palta.* Ecology, 36(2): 169-175.

Ray, P., Singh, S.B. and Sehgal, K.L. (1966): *A study of some aspects of the river Ganga and Jamuna at Allahabad (U.P.) in 1958-59.* Proc. Nat. Acad. Sci. India, 36 B(3): 235-464.

Reddy, P.M. and Venkateshwarlu, V. (1987): *Assessment of water quality and pollution in the river Tungbhadra near Kurnool, (A.P.),* J. Environ. Biol., 8(2): 109-119.

Reid, G.K. (1961): *Ecology of inland water and estuaries.* Reinhold Publication Corporation. N.V.

Reynolds, C.S. and Allen, J.E. (1968): *Changes in the phytoplankton of Oak Mere, following the introduction of Baserich water.* Bri. Phycol. Bull. 3(3): 451-462.

Rose, M.T., Sanchez-Bayo, F., Crossan, A.N. and Kennedy, I.R. (2006): *Pesticide removal from cotton farm tailwater by a pilot scale ponded wetland.* Chemosphere: 63 (11):1849-58.

Ruggiero, A., Cereghino, R., Figuerola, J., Marty, P. and Angelibert, S. (2008). *Farm ponds make a contribution to the biodiversity of aquatic insects in a French agricultural landscape.* Comptes Rendus Biologies 331(4): 298-308.

Ruttner, F. (1963): *Fundamentals of limnology.* English translation by D.G. Frey and F.E.J. Fry. Toronto University Press, Canada. 307 pp (first Germen edition in 1940.)

Sahai, R. and Sinha, A.B. (1969): *Investigations on bio-ecology of inland waters of Gorakhpur (U.P.)., India I. Limnology of Ramgarh Lake.* Hydrobiologia. 34: 433-447.

Sakthivel, V. and Hitesh Shingadia (2001): *Physico-chemical characteristics of Juhu waters before and after Durga Pooja-* Paper presented at National Workshop on Basic Sciences and Fisheries, September 2001.

Sangu, R.P.S. and Sharma, K.D. (1985): *Studies on water pollution on Yamuna river in Agra.* Indian J. Environ. Health., 27(3):257-261.

Scheffer, M., van Geest, G. J., Zimmer, K., Jeppesen, E., Sondergaard, M., Butler, M. G., Hanson, M. A., Declerck, S. and De Meester, L. (2006). *Small habitat size and isolation can promote species richness: second-order effects on biodiversity in shallow lakes and ponds.* Oikos 112(1): 227-231.

Schwerdtfeger, F. (1975): *Synololologie-Parey.* Homburg Berlin 1-451.

Sehgal, K.L. (1983): *Plankters copepods of fresh water Ecosystems.* Environmental Science Series. Inter-print, New Delhi. 169 pp.

Sengar, R.M.S., Sharma, K.D. and Pathak, P.D. (1985): *Studies on distribution of algal flora in polluted and non polluted regions in Yamuna river at Agra (U.P.), J. Indian Bot. Soc.,* 64: 365-376.

Shaikh, Nisar and Yeragi, S.G. (2003): *Biodiversity of plankters in Tansa river of Thane district, Maharashtra.,* J.Nat. Con., 15 (2): 361-366.

Sharma, A.L.N. and Pattanick, P.K. (1985): *Ecological Studies on zooplankters of freshwater ponds in and around Bhubaneshwar.* J. Environ. Biol., 6(4): 245-256.

Sharma, L.L. (1980): *Some limnological aspects of Udaipur water in comparison to selected waters of Rajasthan,* Ph.D. Thesis Univ. of Udaipur.

Sharma, M.S. and Durve, V.S. (1985): *Trophic status and fisheries potential of Rajasthan waters.* Proc. Nat. Sympos. Evalu. Environ., (SPL Vol. Geobios), pp 180-186.

Sharma, N. (1978): *A note on fresh water cladocerans from West Bengal, Bangladesh.* J. Zool., 6: 149-151.

Sharma, R.C. (1985): *Seasonal abundance of phytoplankters in the Bhairathi River,* Garhwal Himalaya. Indian J. Ecol., 12(1): 157-160.

Sharma, S.D. and Pandey, K.S. (1998): *Pollution status of Ramganga river at Moradabad,* Poll. Res., 17(2); 201-209.

Sharma, S.K., Mudgal, L.K. and Balsare, D.K. (2002): *Biodiversity benthic communities of Sirpur Lake, Indore.* Him. J. Env. Zool., Vol. 16(2) pp 231-233.

Shreenivasan, A. (1970): *Limnology of a tropical impoundment: A comparative study of the major reservoirs in Madras state, India,* Hydrobiologia, 36 (3 and4): 443-469.

Shreenivasan, A. (1974): *Limnology features of a tropical impoundment, Bhavanisagar reservoir (Tamil Nadu), India,* Intl. Revue ges., Hydrobiologia, 53(3): 327-342.

Siddiqui, E.N., Singh, N.K. and Afsar, N.R. (1980): *Primary production studies in a freshwater pond at Barauni, Bihar (India).* Comp. Physiol ecol., 5(4): 258-260.

Singh, A.K. and Ahmad, S.H. (1990): *A comparative study of the phytoplankters of the River Ganga and Pond of Patna (BIHAR), India,* J. Indian bot. Soc., 69:153-158.

Singh, D.F. and Rai, M.K. (1988): *Studies on limnology of Bada Talab.* J. Environ. Bio., 9(1): 69-71.

Singh, F.D. (1986): *Relationship between primary productivity and environmental parameters in tropical lakes: Statistical analysis.* Poll. Res. 5 (3&4): 103-110.

Singh, H.P., Chaudhry, M. and Kalekaur, V. (1982): *Seasonal and diurnal changes in physico-chemcial features of the river Brahamputra at Guwahati,* Ind. J. Zool., 2(82);77-84.

Singh, M. (1968): *Phytoplankters and water temperature silicon and pH in a lake in Delhi.* Phykos, 7:120-128.

Singh, M. and Swarup, K.U. (1979): *Limnological studies of Suraha Lake (Ballia). II. The Periodicity of Phytoplankters.* J. Indian Bot. Soc., 58: 319-329.

Singh, R.K. and Desai, V.R. (1980): *Limnological observations on Rihand reservoir. III Primary Productivity.* J. Inland Soc. India, Vol. 12, No.2: 63-68.

Singh, R.N. (1965): *Limnological relations of Indian inland water blooms* Proc. Int. Assoc. Limnol. 12: 831-836.

Solimini, AG, Della Bella V, Bazzanti M. (2005). *Macroinvertebrate size spectra of Mediterranean ponds with differing hydroperiod length.* Aquatic Conservation: Marine and Freshwater Ecosystems 15: 601–611.

Sondergaard, M. Brunn, L., Lauridsen, T., Jeppesen, E., and Madsen, T.V., (1996): *The impact of grazing waterfowl on submerged macrophytes in Situ experiments in a shallow eutrophic lake.* Aquat. Bot. 53, 73-84.

Srivastava, N., Agrawal, M. and Tyagi, A. (2003): *Study of physico-chemical characteristics of water bodies around Jaipur.* J. Environ. Biol., 24, 177-180.

Stoianov, I., Chapra, S. and Maksimovic, C. (2000). *A framework linking urban park land use with pond water quality.* Urban Water 2: 47 - 62.

Stross, R.G., Miller, M.C. and Daley, R.J. (1980): *Zooplankters communities Life cycles and production. J.E. Hobbie (ed.), Limnology of Tundra Ponds.* Scademic Press, N. York. 251-296.

Tan, H, Hu, Y. and Mei, Z. (2002): *Main factors influenceing the growth of planktonic bacteria in fish pond ecosystem.* Ying Yong Sheng Tai Xue Bao. 13 (1): 95-100.

Thresh, J.C., Sucbling, E.V., and Beale, J.F. (1944): *The examination of water supplies.* Ed. 6 by Taylor, E.W.

Tirath, R. and Bhargawa, D.S. (1982): *Water Quality in Gomati river at Lucknow (U.P.),* J. Ind. water works Assoc., 14:299-304.

Trivedi, R.C. (1979): *Pollution studies of chambal river and surroundings due to Nagda industrial complex.* Ph.D. Thesis. School of studies in Botany, Vikram University, Ujjain.

Trivedi, R.K. and Goel, P.K. (1984): *Chemical and biological methods for water pollution studies.* Environmental publications, Karad:1-250 pp.

Trivedi, R.K., Garud, J.M. and Goel, P.K. (1983): *Limnology of five lakes in Kolhapur with reference to human activity.* Environ. And Ecol., 1: 41-43.

Tucker, D.S. (1958): *The distribution of freshwater invertebrates in ponds in relation to annual fluctuations in the chemical compositions of the water.* J. Anim. Ecol. 27:105-123.

Tundisi, M.T. and Tundisi, J.G. (1976): *Plankters studies in a lacustrine environment. I. preliminary data on zooplankters ecology of Brca reservoir.* Oecologia, (Berl) 25: 265-270.

Unni, K.S. (1982); *Limnological studies on Sumpna reservoir, Betul, M.P.,* Proc. Nat. Sci Acad., India 52(B), IV.

Usha, R., Ramalingam, K. and Ramjam, U.D. Bharathi (2006): *Freshwater lakes – A potential source for aquaculture activities - A model study on Perumal lake, Cuddalore, Tamil Nadu.* J. Environ. Biol., 27, 713-722.

Vasisth, H.S. and Battish, S.K., (1971): *The rotifer fauna of north India.* Res. Bull. (N.S.) Punjab University 22 (1 and 2): 179-188.

Vasisth, H.S. and Sharma, B.K. (1975): *Ecology of a tropical urban pond in Ambala city of Haryana State.* Ind. J. Ecol., 2(1): 79-86.

Venkateshwarlu, T. (1969): *An ecological study of the algae of the river Moosi Hyderabad (India) with special reference to the effect of Microsystis aeruginosa (Kutza) bloom on fish production and suggestions for the improvement of the tank fishery.* Ichthyologica, 3: 37-46.

Verma, M. (1967): *Diurnal variation in a fish pond in Seoni, India* Hydrobiologica, 30: 129-137

Verma, P.K. and J.S. Datta Munshi (1987): *Plankters Community structure of Badua Reservoir, Bhagalpur (Bihar).* Trop. Ecol. 28: 200-207.

Verma, S.R. and G.R. Shukla (1970): *The physic-chemical condition of Kamla Nehru Tank, Muzzafarnagar (U.P.) in relation to biological productivity.* Environ. Hlth.12:110-128.

Verma, S.R. and G.R. Shukla, (1969): *Pollution in a perennial stream Khala, by the sugar factory effluent nearLaksar (Dist. Saharanpur) U.P.* India Env. Hlth. 11: 145-162.

Verma, S.R., Sharma, P., Tyagi, A., Rani, S., Gupta, A.K. and Dalela, R.C. (1984): *Pollution and saprobic status of eastern Kalinadi.* Limnologica, 15 (1): 69-133.

Vermonden, K., Leuven, R., van der Velde, G., van Katwijk, M. M., Roelofs, J. G. M. and Hendriks, A. J. (2009). *Urban drainage systems: An undervalued habitat for aquatic macroinvertebrates.* Biological Conservation 142(5): 1105-1115.

Vollenweider, R.A. (1984): *The scientific basis of the lakes and streams eutrophication with particular reference to the nitrogen and phosphorus as eutrophication factors.* Tech. Rept. OECD Paris, DAS/DSI, 68(27)1:182.

Waglenska, T, Dylinska, L.B., Kartin, T.E. and Spoiska, I, (1987): *Ecological pollution,* 173-208.

Ward, H.B. and Whiple, G.C. (1959): *Fresh Water Biology* 2nd edition. John Wiley and Sons. INC. N. York, London.

Ward, R.W. and Seibert, H.C. (1930): *Investigation into the limnology of Dow lake.* Athens Country. *Ohio J. Sci.,* 63(3): 128-144.

Welch, P.S. (1935): *Limnology.* Mcgraw Hill, Book Co. Inc., New York, 471.

Welch, P.S. (1948): *Limnological Methods.* The Blakiston. Co. Philadelphia, p. 1-381.

Welch, Paul S. (1952): *Limnology, 2nd ed.* McGraw Hill Book Co. Philadelphia, 1-381.

Wetzel, R.C. (1975): *Limnology,* W.B. Saunders Company. Philadelphia. 743.

Wetzel, R.J., Likens, G.E. (1979): *Limnological analysis,* Philadelphia, W.B. Saunders 357.

Whitton, B.A. (1985): *Biological monitoring of heavy metal in flowing waters.,* Symp. Biomonitoring state of Environment, 50-55.

Williams, P., Whitfield, M., Biggs, J., Bray, S., Fox, G., Nicolet, P., Sear, D. (2004). *Comparative biodiversity of rivers, streams, ditches and ponds in an agricultural landscape in Southern England.* Biological Conservation 115: 329–341.

Williams, P. J., Biggs, J., Barr, C. J., Cummins, C. P., Gillespie, M. K., Rich, T. C. G., Baker, A., Baker, J., Beesley, J. and Corfield, A., *et al.* (1998). *Lowland Pond Survey 1996,* Department of the Environment, Transport and the Regions: London.

Wlakeel, S.K. and Waby, S.P. (1970): *Hydrography and chemistry of lakes Manzallah, Egypto Arch.* Hydrobiol., 67 (2): 173-200.

Wood, P. J. and Barker, S. (2000). *Old industrial mill ponds: a neglected ecological resource.* Applied Geography 20(1): 65-81.

Wu, Q.L. and Hahn, M.W., (2006): Differences in structure and dynamics of polynucleobacter communities in a temperature and subtropical lake, revealed at three phylogenetic level. *FEMS Microbiol. Ecol.*, 57(1): 67-79.

Zafar, A.R., (1966): Limnology of the Hussain Sagar Lake, Hyderabad, India. *Phykos.*, 5(1&2): 115-126.

Index

A

Aquatic flora, 96

Aspergillus flavus, 109, 119

B

Biological examination, 30

Brilliant Green Lactose Bile (BGLB), 31

C

Candida albicans, 109, 112, 119, 122

Catla-catla, 148

Central Building Research Institute, 7

Chanda-nama, 148

Channa-stratius, 148

Cirrhinus-nrigala, 148

Cladocera, 95

Cyprinus-carpio, 148

D

Discussion, 123-148

F

Fish diversity, 30

Fisheries Department in Paniyala, 95

Fishes, 95-96

Fusarium, 110, 112, 119, 122

G

Ganga canal, 7

H

Hypophthalmichthys-molitrix, 148

I

Introduction, 1-6

L

Labeo-rohita, 148

Limnology, 2

M

Materials and methods, 19-33

- biological examination, 30
- chemicals, 20
- collection of plankton, 30
- fish diversity, 30
- instruments, 20
- methodology, 24-30
- microbiological parameters, 30-33
- plankton, 30

Microbial and Planktonic Ecology of Paniyala State Fish Pond at Roorkee, 149

N

Notopterus-notopterus, 148

O

Ostracods sp., 95

P

Paniyala Fish Pond, 6, 95, 123, 148, 149

Parameters for assessing pond water quality, 34-122

- aquatic flora, 96
- biological parameters, 67
- chemical characteristics, 44-67
- fishes, 95-96
- microbiological characteristics, 96
- physical characteristics, 34-44
- plankters, 67-95

Penicillium spp., 109, 112, 119, 122

Physico-geographical features, 7-11

Physico-geographical features: sampling sites, 8-11

Plankters, 67-95

Plankton, 30

Planktonic Ecology of Paniyala State Fish Pond at Roorkee, 149

Pond water quality, 34-122

R

Red Data Book, 2

Review of literature, 12-18

- Daborn, 13
- Eddy, 12
- Gopal, 14
- Hosmani, 16
- Khan, 13
- Khanna, 16
- Needham, 16

River Ganga, 13

S

Sampling methods, 20

- analytical methods, 23-24
- cleaning and sterilization of glassware, 20-21
- collection of samples, 21-22
- reagent preparation, 22-23

Sampling programme and analysis, 19

State Fish Department, 18

U

U.P. Irrigation Design Organization, 7

United Nations, 2